Springer Undergraduate Mathematics Series

SUMS Readings

Editor-in-Chief

Endre Süli, Oxford, UK

Series Editors

Mark A. J. Chaplain, St Andrews, UK

Angus Macintyre, Edinburgh, UK

Shahn Majid, London, UK

Nicole Snashall, Leicester, UK

Michael R. Tehranchi, Cambridge, UK

SUMS Readings is a collection of books that provides students with opportunities to deepen understanding and broaden horizons. Aimed mainly at undergraduates, the series is intended for books that do not fit the classical textbook format, from leisurely-yet-rigorous introductions to topics of wide interest, to presentations of specialised topics that are not commonly taught. Its books may be read in parallel with undergraduate studies, as supplementary reading for specific courses, background reading for undergraduate projects, or out of sheer intellectual curiosity. The emphasis of the series is on novelty, accessibility and clarity of exposition, as well as self-study with easy-to-follow examples and solved exercises.

David Nacin

The Group Theory
Puzzle Book

 Springer

David Nacin 🆔
Department of Mathematical Sciences
William Paterson University
Wayne, NJ, USA

ISSN 1615-2085   ISSN 2197-4144   (electronic)
Springer Undergraduate Mathematics Series
ISSN 2730-5813   ISSN 2730-5821   (electronic)
SUMS Readings
ISBN 978-3-031-70090-3   ISBN 978-3-031-70091-0   (eBook)
https://doi.org/10.1007/978-3-031-70091-0

Mathematics Subject Classification: 20-01, 00A08, 00A05

This Springer imprint is published by the registered company Springer Nature Switzerland AG
The registered company address is: Gewerbestrasse 11, 6330 Cham, Switzerland

If disposing of this product, please recycle the paper.

Preface

To students in an introductory course, groups can sometimes seem like mystical abstractions that mathematicians enjoy proving theorems about. Concrete examples and computations are sometimes provided, but how can we gamify these computations to make them more engaging and fun? What if there were a type of game that not only addressed this issue but also motivated the basic definition of group, as well as later ideas from a group theory course?

This book is designed to introduce readers to a new type of puzzle: one using neither letters nor numbers in the traditional sense, but instead using the elements of a mathematical structure called a finite group. We present a collection of these group theory puzzles for everyone to enjoy, from those being introduced to groups for the very first time, to those well familiar with all the groups mentioned in this work. No prior math knowledge is required. We include brief introductions to each family of groups we will be using, fully explaining the basic mechanics of each group product, and thus providing everything needed to try and eventually master the puzzles presented here. We leave the beautiful proofs contained in a typical group theory class to the many excellent textbooks that already exist, which allows us to include as many puzzles as possible and keep the majority of pages here useful to all readers [1, 2, 5–7, 20].

The later sections of this book are devoted to additional techniques that can be used to help tackle the problems within. These can be read at any time and apply to all the group theory puzzles presented here. We place these optional sections near the end to promote active learning by encouraging the reader to discover some of these approaches independently. We make no claims that the tricks listed in that chapter are comprehensive, and we encourage readers to discover methods beyond those listed, perhaps even uncovering tricks unknown to anyone so far.

Because larger puzzles can take a significant amount of time to complete, we have chosen to focus on groups with eight or fewer elements. Puzzles can still be made quite challenging without resorting to large sizes. We simply note here that there is nothing preventing the creation of similar puzzles over larger groups.

This book can be used as a supplement to an abstract algebra class for instructors who want students to have more hands-on practice with group elements, providing concreteness alongside the usual abstraction. It can also be used by individuals studying on their own who want exposure to many examples of common groups before moving on to a proof-based text. Finally, this book can be enjoyed by those already familiar with groups who simply want to tackle some clever puzzles and possibly discover new facts about operations they already know.

Wayne, NJ, USA   David Nacin

Acknowledgements

I would like to thank Michael Beals, the analyst who first introduced me to group theory, and Robert Wilson, the algebraist who guided me the rest of the way.

My interest in puzzles was sparked by Brian Hayes' article, "KenKen-friendly Numbers," on his bit-player blog, where he introduced a puzzle involving complex numbers. Had he not left the uniqueness of the solution as an open problem for his readers, I may have never discovered my love for strange puzzles, and this book might not exist.

I am also deeply grateful to my wife, Erin, for her constant kindness and support throughout this project and all others.

Contents

Integer Calcudoku

1

This is a book of and about puzzles. Most paper puzzles involve filling in things, so we begin by discussing those things we fill. Mathematicians use the word *set* to refer to any collection of objects. For example, in a Sudoku puzzle we fill in the individual squares, which we call *cells*, with entries from the set of numbers from one to nine. In English crossword puzzles we fill in the cells with the elements of the set 'A' through 'Z'. Crosswords are popular in many other languages, and since some of those languages use a different alphabet, they require a different set of characters.

Not all sets are defined through verbal descriptions. Small sets can be described by listing all their elements within curly brackets. For example, instead of saying the "the set of numbers from one to nine" as we did before, we might simply write $\{1, 2, 3, 4, 5, 6, 7, 8, 9\}$. We could also use ellipsis, the series of three dots symbolizing intentional omission, if there are more elements than we wish to write. This should only be done once enough terms are written that it is clear that the set is understood, allowing us to express our set using something like $\{1, 2, 3, \cdots, 8, 9\}$ or $\{1, 2, \cdots, 9\}$.[1]

A *Latin square* over a set is an array of entries from that set, where each element of the set appears exactly once in each row and each column. For example, Fig. 1.1 shows a Latin square over the set of numbers one through seven, and Fig. 1.2 shows one over the set of letters in the first half of the English alphabet.

These are specific examples, each representing just one out of many possible choices. One of the first questions one might ask is: How many such choices are there?

When counting the number of possible Latin squares, the specific symbols used do not affect the count; only the size of the set matters. Regardless, the general answer to exactly how many Latin squares exist remains unknown. For small

---

[1] Of course, this brings up many issues, both psychological and philosophical, about, when or whether it is at all possible for us to know when a pattern is understood. Such questions are beyond the scope of this humble puzzle book.

© The Author(s), under exclusive license to Springer Nature Switzerland AG 2024
D. Nacin, *The Group Theory Puzzle Book*, SUMS Readings,
https://doi.org/10.1007/978-3-031-70091-0_1

Fig. 1.1 A Latin square over
the set of numbers one
through seven

$$\begin{bmatrix} 3 & 1 & 2 & 5 & 6 & 4 & 7 \\ 2 & 6 & 5 & 4 & 7 & 3 & 1 \\ 7 & 4 & 3 & 6 & 1 & 5 & 2 \\ 4 & 7 & 1 & 3 & 2 & 6 & 5 \\ 5 & 3 & 7 & 1 & 4 & 2 & 6 \\ 1 & 5 & 6 & 2 & 3 & 7 & 4 \\ 6 & 2 & 4 & 7 & 5 & 1 & 3 \end{bmatrix}$$

Fig. 1.2 A Latin square over
the set of letters in the first
half of the English alphabet

$$\begin{bmatrix} D & H & J & K & G & B & M & A & L & E & I & C & F \\ J & I & B & C & M & E & H & F & A & G & L & K & D \\ I & D & G & A & J & L & C & H & K & B & F & E & M \\ K & E & I & B & H & G & D & L & F & C & J & M & A \\ B & M & D & G & E & J & L & K & C & I & A & F & H \\ A & B & H & E & C & D & I & G & J & F & M & L & K \\ C & G & L & D & A & K & F & M & H & J & B & I & E \\ M & F & A & L & K & H & J & B & E & D & C & G & I \\ L & K & C & F & D & I & G & J & M & H & E & A & B \\ F & J & E & I & B & A & K & C & D & M & G & H & L \\ E & L & M & J & F & C & A & D & I & K & H & B & G \\ H & A & K & M & I & F & B & E & G & L & D & J & C \\ G & C & F & H & L & M & E & I & B & A & K & D & J \end{bmatrix}$$

sets with eleven elements or fewer, we've been able to count them. There's one Latin square of size one, two of size two, twelve of size three, and five hundred and seventy-six of size four. These numbers increase rapidly without any efficient formula. For any size greater than or equal to twelve, we have no exact number, though estimates are obtainable. Even for squares of size eleven, the exact number was only discovered as recently as 2005[2] [14].

Many number puzzles, including the extremely popular Sudoku, are designed around the premise of completing a Latin square. In Sudoku puzzles, an additional constraint involves three-by-three square regions called cages, which also must contain each element exactly once. Some cells are initially revealed to ensure a unique solution. When it comes to solving such puzzles, the symbols used do not matter at all. Any set of nine easily distinguishable characters works just as well as any other. With our puzzles in this book, we wish to head in a different direction.

In the year 2004, a schoolteacher named Tetsuya Miyamoto designed the first KenKen puzzles in an attempt to make learning and practicing basic arithmetic operations more fun for his students. Though his puzzles are suitable for children, they can also be made challenging enough for adults, and his puzzle quickly spread around the globe. It is now featured regularly in many famous publications including the New York Times. Miyamoto still teaches in both New York and Japan and follows a philosophy he calls "The Art of Teaching Without Teaching."

---

[2] To give a scale of the sizes involved, with a set of eleven elements, there are 776,966,836,171,770,144,107,444,346,734,230,682,311,065,600,000 ways to make a Latin square.

This is an excellent example of *gamification*, the act of using of game mechanics in order to engage and motivate people towards serious goals. Such techniques are becoming more common in education, and though Miyamoto's puzzles may provide an excellent way to introduce arithmetical operations, there really is no reason whatsoever to restrict our operations to those over the integers. We will show that there are many reasons why these puzzles work even better for different operations over different mathematical structures. For right now, we will provide some examples in a more standard setting: puzzles over the set $\{1, 2, 3, \cdots, n\}$ with a single arithmetic operation. In this chapter, we refer to these as *Calcudoku puzzles*, a commonly used name for these puzzles that avoids copyrighted terms.

As with Sudoku, the goal of each of each size n-by-n Calcudoku puzzle is to fill in the cells with the numbers $\{1, 2, 3, \cdots, n\}$ so that the final result is a Latin square. In addition, the grid is partitioned into heavily outlined regions that we will call *cages*. Some or all of the cages have clues in the form of a target number. When the operation is applied to each of the values in the cells of that cage, the output must reach that target number. For an example, consider the following Calcudoku puzzle over the set $\{1, 2, 3, 4\}$ where the operation is addition. Notice how in the solution, the sum of the entries in each of the cages equals the clue for that cage (Fig. 1.3).

There are many different ways to solve a puzzle like this. One approach is to start in the bottom left cage, where a clue of four means that those two cells must contain the numbers 1 and 3 in some order, since we cannot have duplicates in the same row. Consequently, the bottom right cage must contain the numbers 4 and 2, and as the entries in that cage add to 10, it must contain a second 4. There is only one way to place two 4's in that cage without having two in the same row or column, so we can fill in the cells as shown in Fig. 1.4.

The clue of 5 in the top right cage indicates that those cells contain either 2 and 3, or 1 and 4. Since there is no way to place a 4 there, we must use the numbers 2 and 3, which we can only do in one way. Also, since we cannot have two 3's in the same column, the six in the top left indicates a 2 and 4, which can only be positioned as shown in Fig. 1.5.

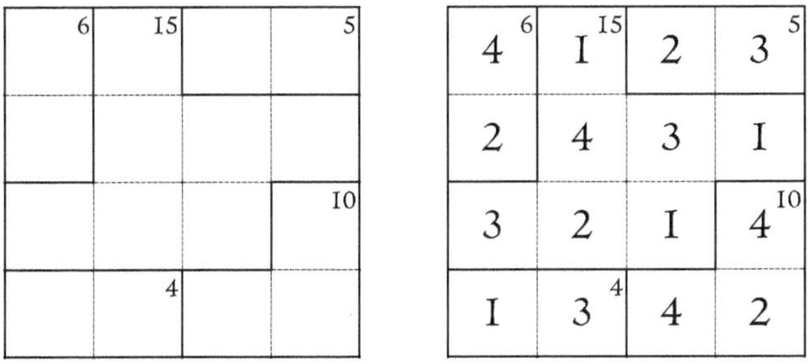

Fig. 1.3 An additive calcudoku puzzle and its solution over $\{1, 2, 3, 4\}$

Fig. 1.4 The only possibility
for the lower right cage

Fig. 1.5 Filling in two upper
cages

Fig. 1.6 Placing the 1's in
the only way possible

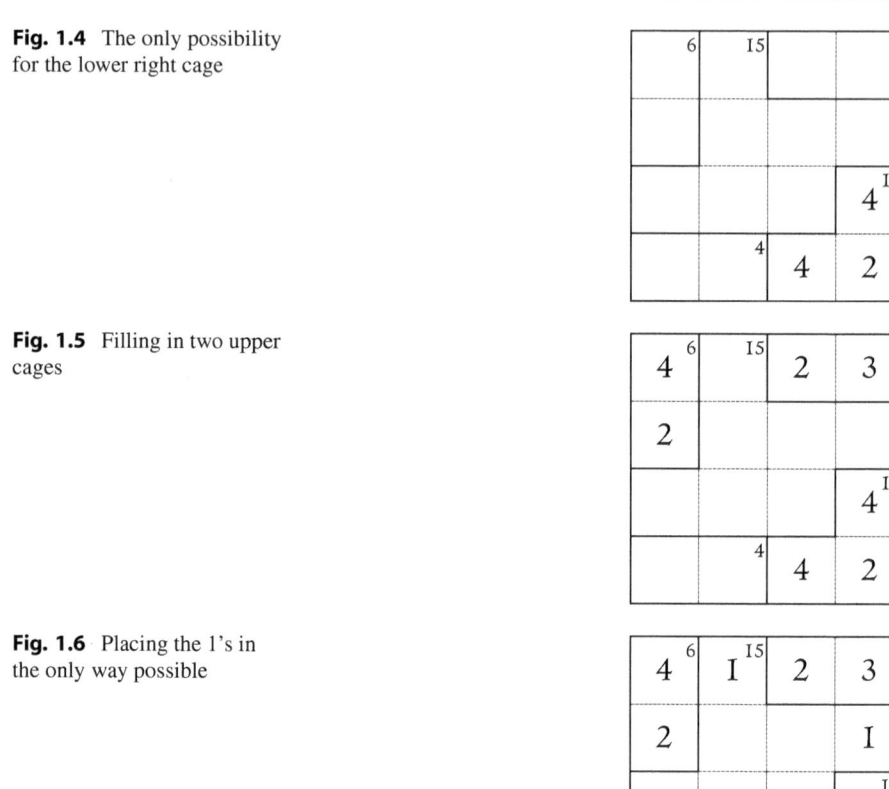

We still have to ensure a single 1 appears in every row and column. Due to the
currently occupied cells, this can only be done in the way shown in Fig. 1.6, and
from there the Latin square condition should be enough for us to quickly arrive at
the solution.

Now is a good time to start attempting some puzzles (Figs. 1.7, 1.8, 1.9, 1.10,
1.11, 1.12, 1.13, 1.14, 1.15, and 1.16). Each puzzle in the upcoming pages features
a single operation. For puzzles with single-cell cages, the clue is simply the value of
that cell, regardless of the operation used.

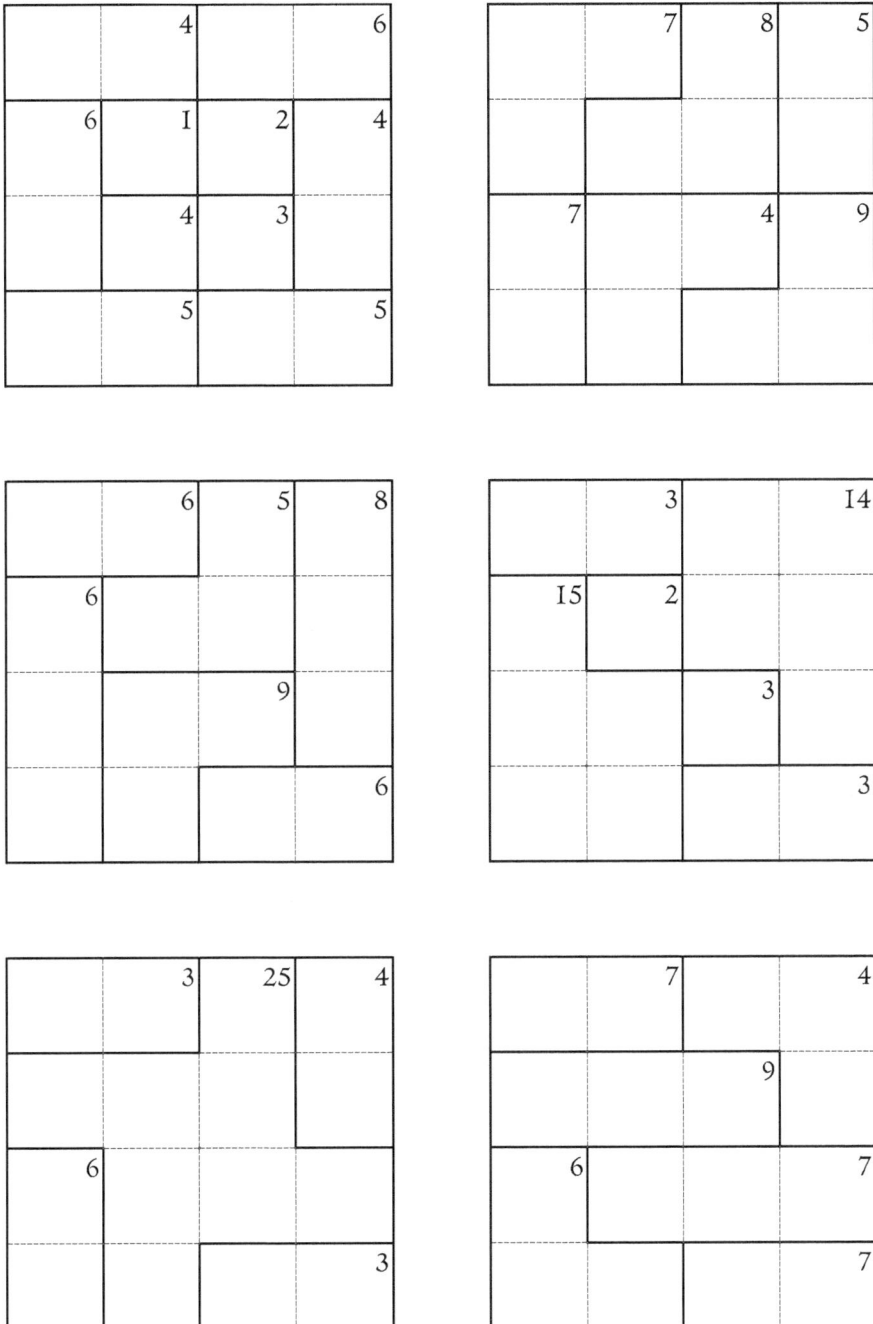

Fig. 1.7 Six additive puzzles over the integer set {1, 2, 3, 4}

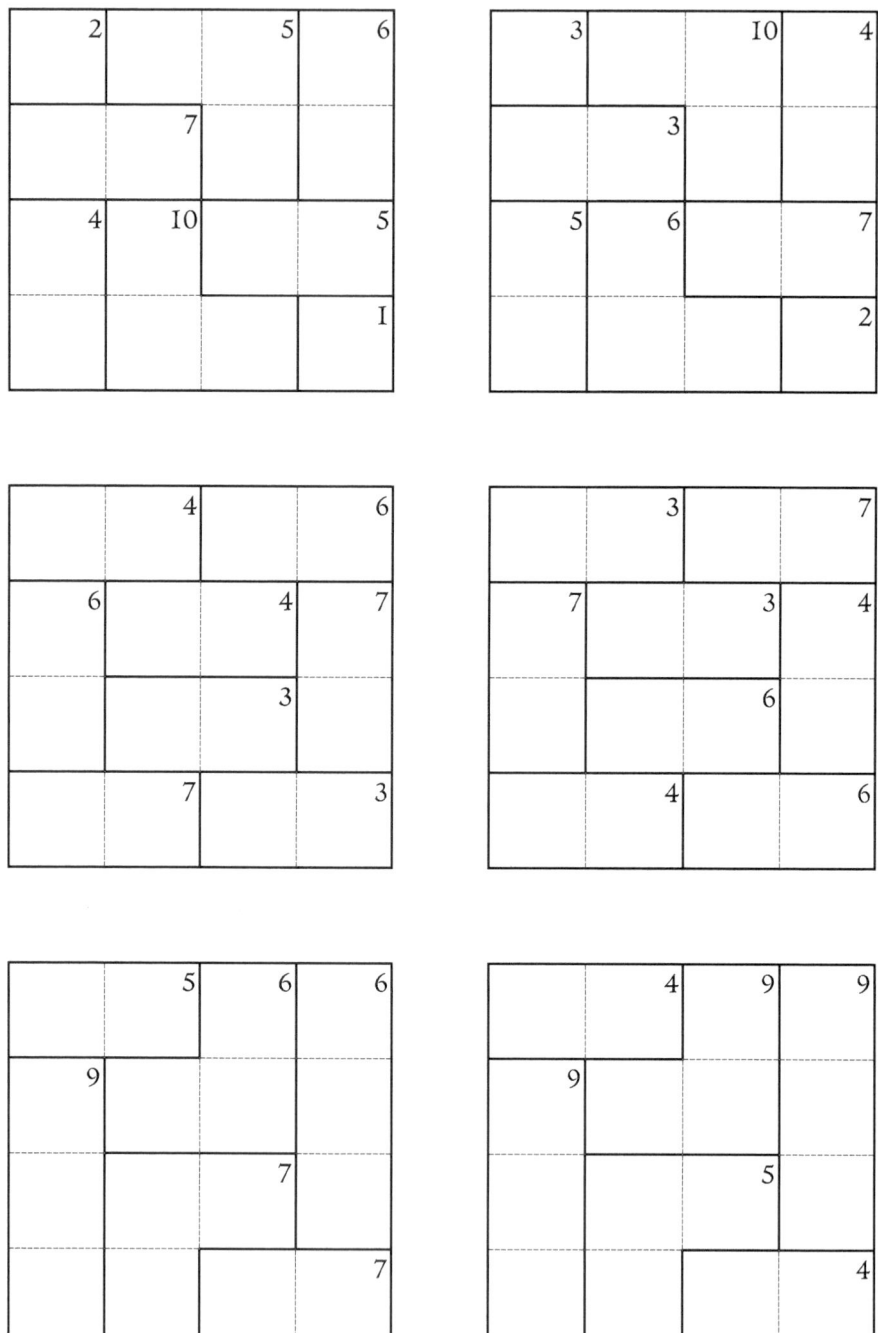

Fig. 1.8 Six additive puzzles over the integer set {1, 2, 3, 4}

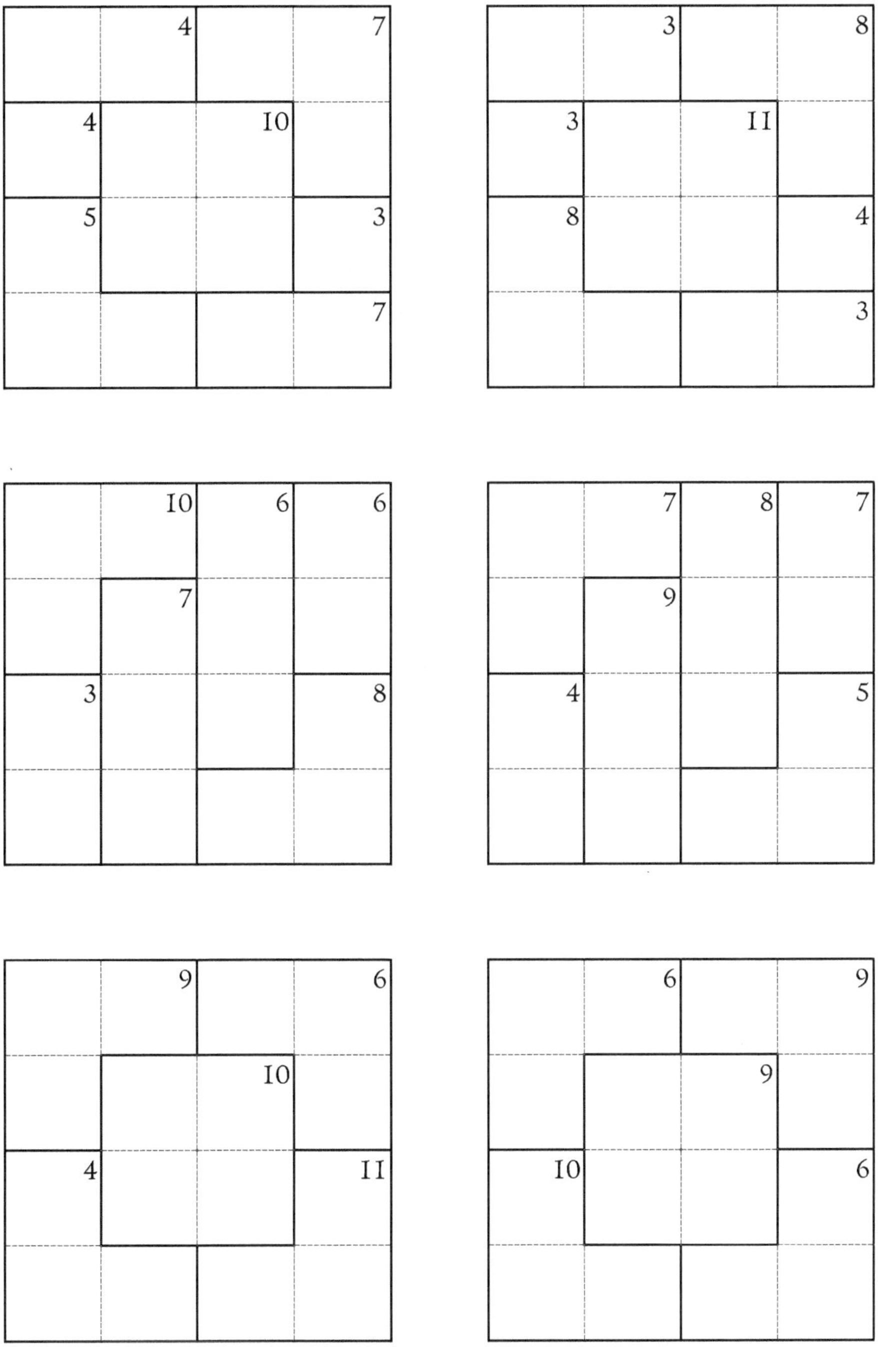

Fig. 1.9 Six additive puzzles over the integer set $\{1, 2, 3, 4\}$

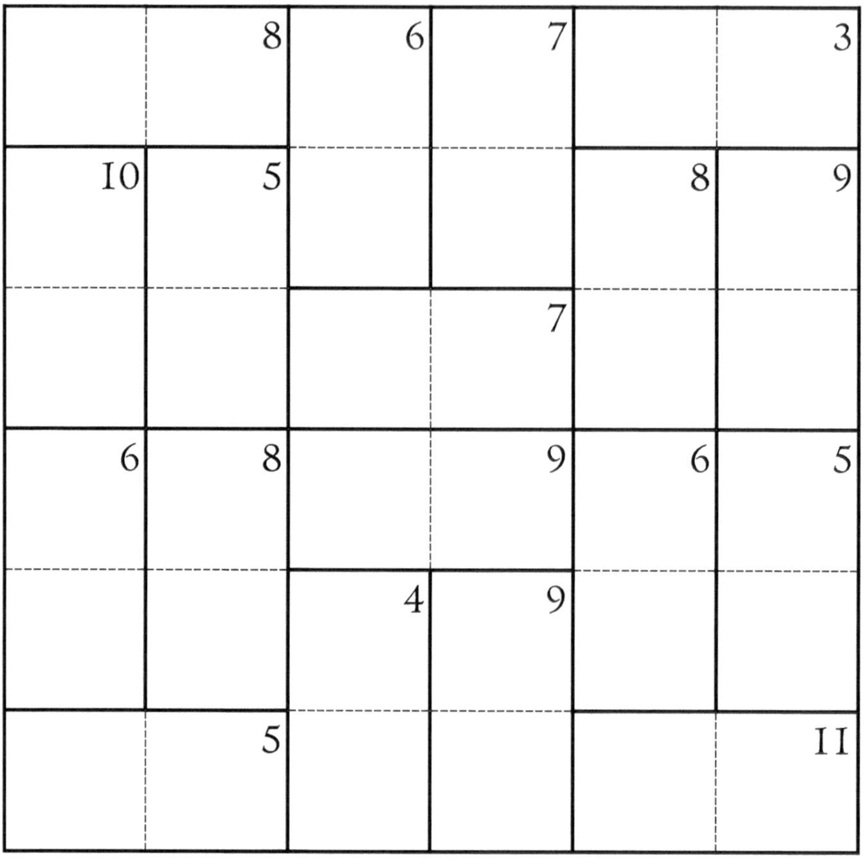

	8	6	7		3
10	5			8	9
		7			
6	8		9	6	5
		4	9		
	5				11

Fig. 1.10 An additive puzzle over the integer set $\{1, 2, 3, 4, 5, 6\}$

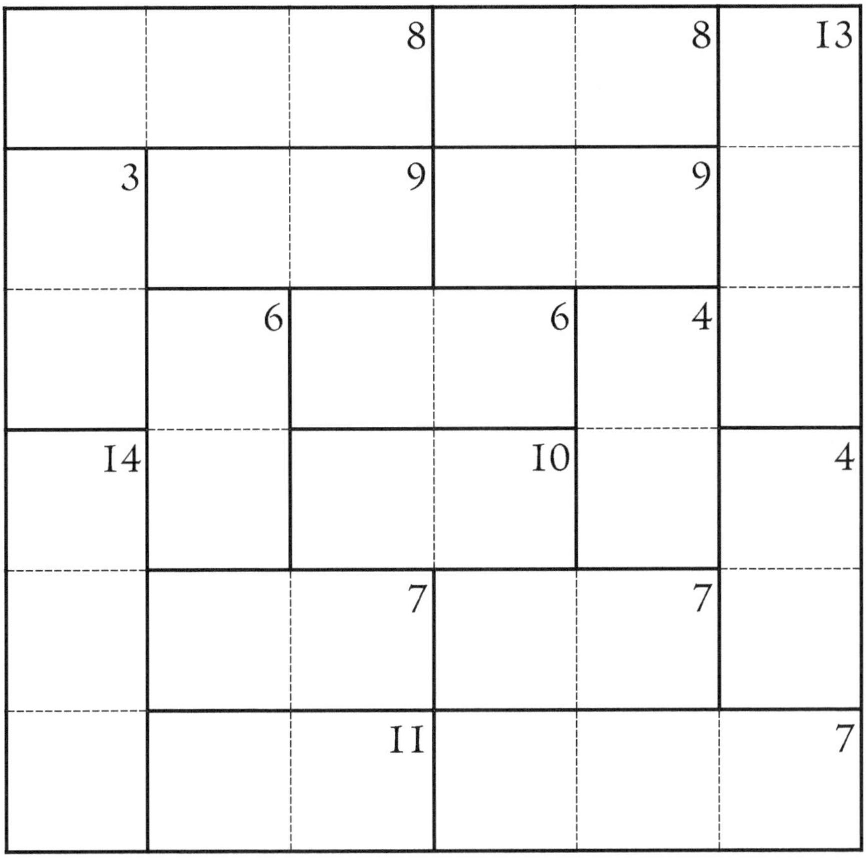

Fig. 1.11 An additive puzzle over the integer set $\{1, 2, 3, 4, 5, 6\}$

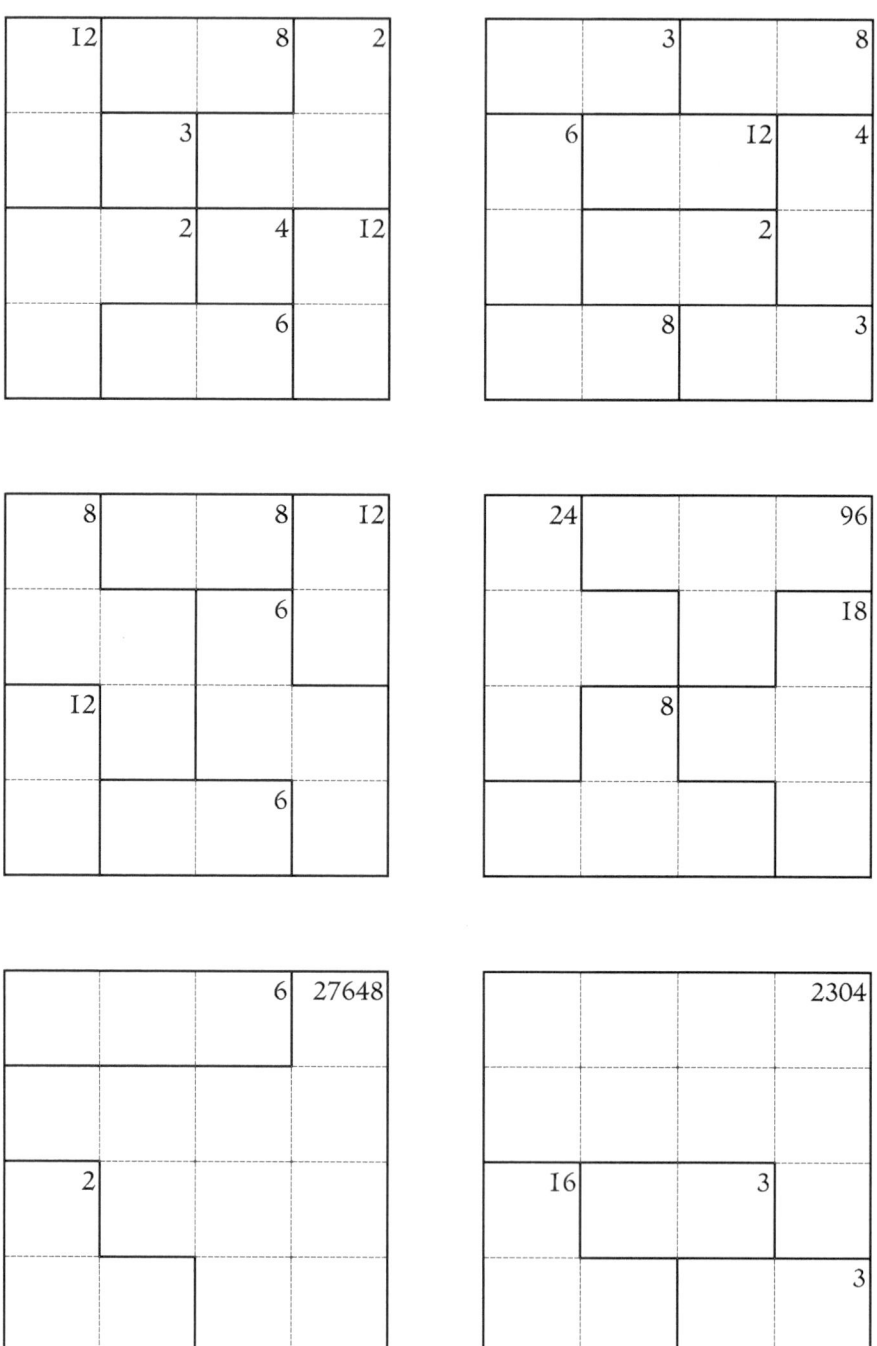

Fig. 1.12 Six multiplicative puzzles over the integer set {1, 2, 3, 4}

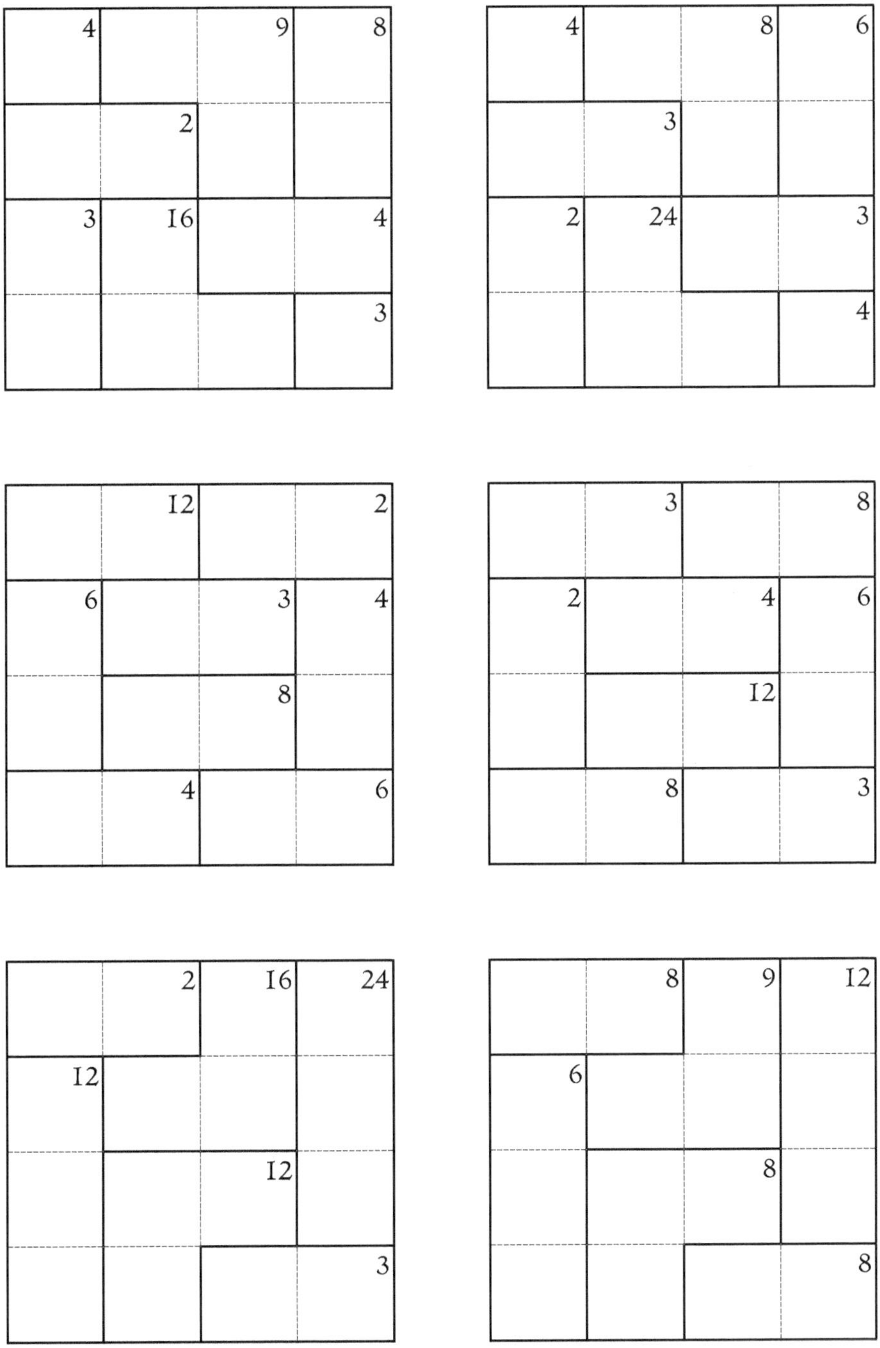

Fig. 1.13 Six multiplicative puzzles over the integer set $\{1, 2, 3, 4\}$

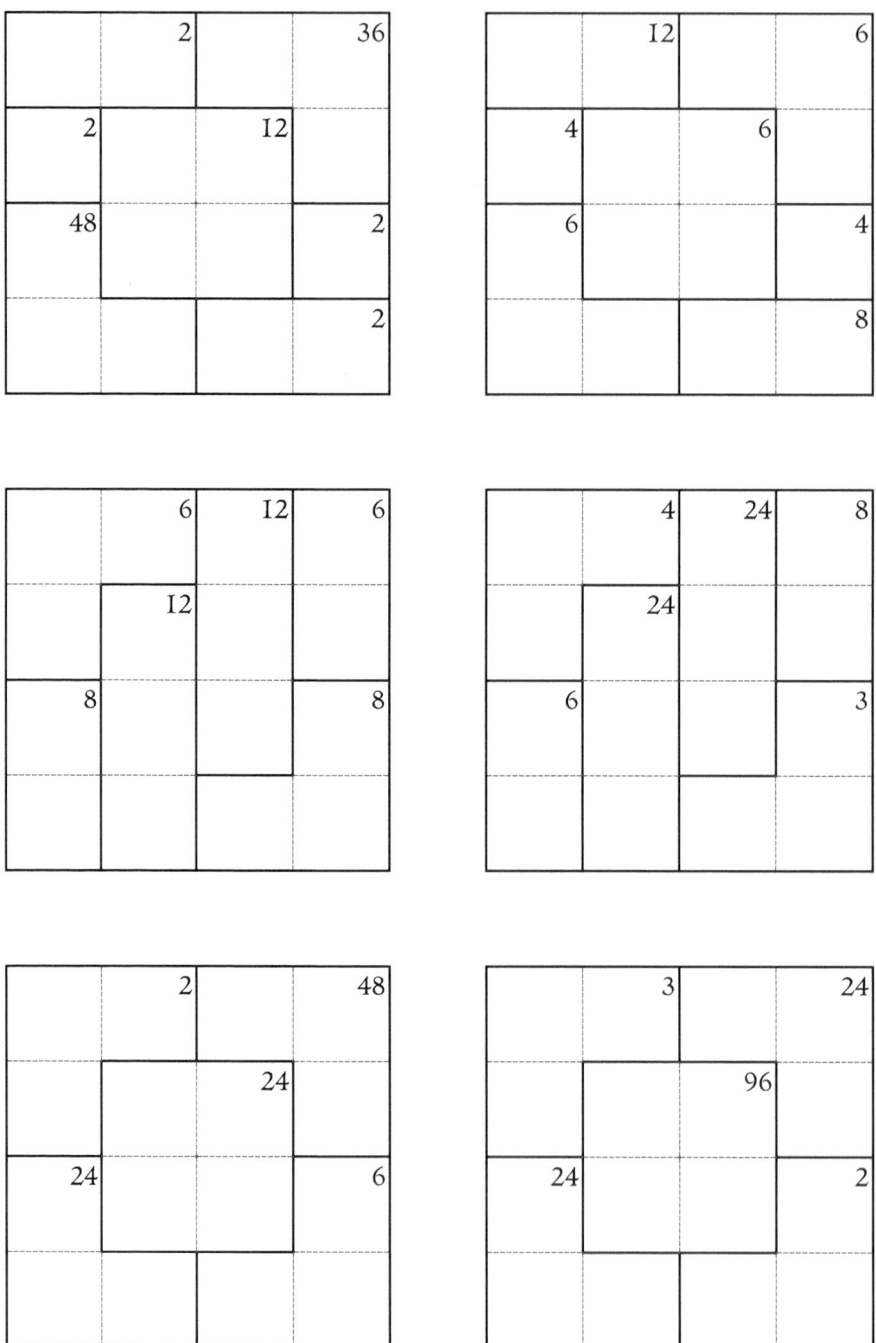

Fig. 1.14 Six multiplicative puzzles over the integer set {1, 2, 3, 4}

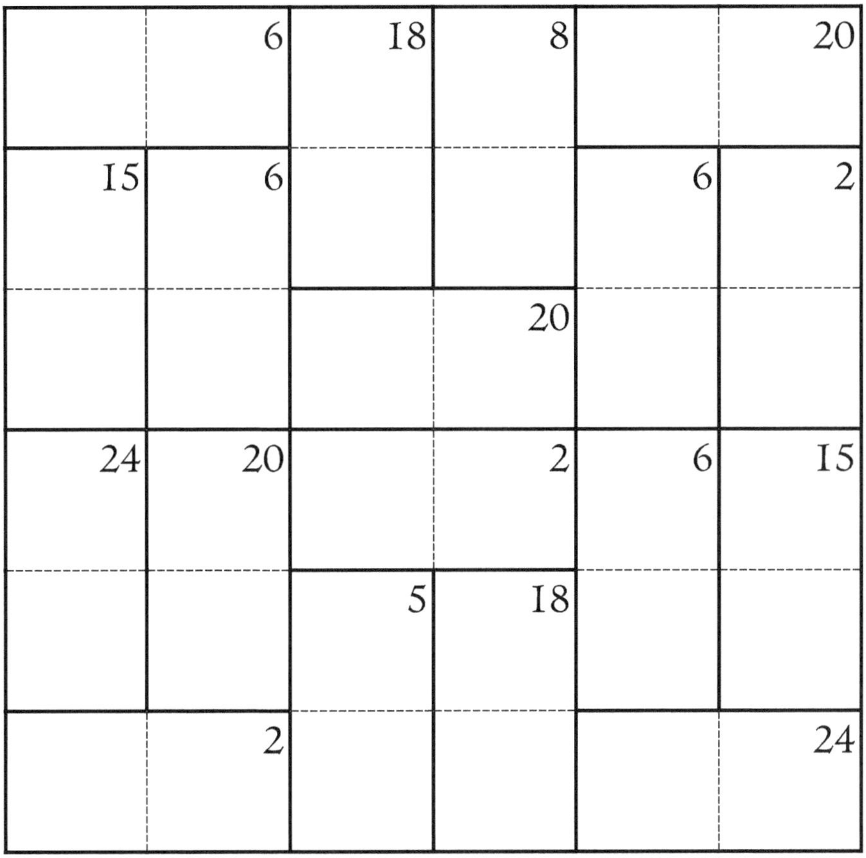

	6	18	8		20
15	6			6	2
			20		
24	20		2	6	15
		5	18		
	2				24

Fig. 1.15 A multiplicative puzzle over integer set $\{1, 2, 3, 4, 5, 6\}$

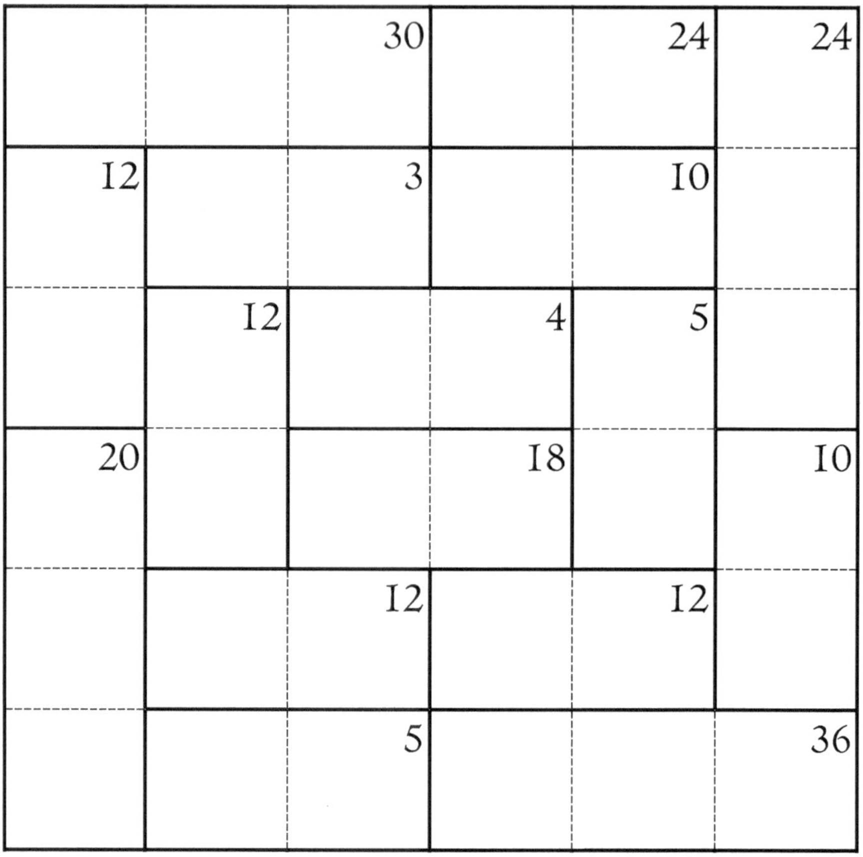

Fig. 1.16 A multiplicative puzzle over integer set $\{1, 2, 3, 4, 5, 6\}$

Let's consider the smallest nontrivial step in a puzzle: solving for a two-cell cage. Call the numbers we are placing into those cells x and y. After applying the operation to x and y, what we get must be equal to the value of some clue c. If the operation is multiplication, then the possibilities for the numbers in those cells will be all the solutions in x and y to the equation

$$x \times y = c.$$

If the operation is addition, then the possibilities will be the solutions to

$$x + y = c.$$

From the clue and any one number, we can completely determine the value of the other. For example, if c is 6 and we can deduce from other cells in the puzzle that y must be 2, we can quickly find the value of x. Our brains may jump straight to the answer in both the multiplicative and additive situations, but let's try to think about how we can find those answers using the most basic rules of arithmetic.

To solve the equation

$$x \times 2 = 6$$

lets first make sure we agree on what that statement means. It asserts that the quantity $x \times 2$ is identically equal to 6. Thus if we were to multiply $x \times 2$ by $\frac{1}{2}$, and multiply 6 by $\frac{1}{2}$, both outcomes would be the same, since the original numbers were the same to begin with. This is true for any number, but we choose $\frac{1}{2}$ because we see a 2 to the left of the equals sign, and the product of 2 and $\frac{1}{2}$ is particularly nice. So far we now know

$$(x \times 2) \times \frac{1}{2} = 6 \times \frac{1}{2}.$$

The product on the right side equals the number 3. On the left side, we have the product of the product of x and 2, whatever that is, multiplied by the number $\frac{1}{2}$. We'd like to take the product of two and one-half, since that's the reason we brought that particular fraction into the equation to begin with. Instinct will probably tell us that it's okay to group things differently on the left. Whether or not one recalls the terminology, they are likely imagining the *associative rule* for multiplication, which states that for any numbers a, b and c, we have

$$(a \times b) \times c = a \times (b \times c).$$

This associative rule doesn't apply universally to all operations. For example $(3 - 2) - 1$ equals zero, but $3 - (2 - 1)$ equals two. Most readers are probably not shocked by this, and due to years of arithmetic, wouldn't have even attempted to use the associative rule when subtraction was involved. Some readers may wish to think about how many operations they know that satisfy or fail to satisfy the associative rule, but it is enough right now to understand that this rule does hold for both multiplication and addition.

Applying the rule gives us the equation

$$x \times (2 \times \frac{1}{2}) = 3.$$

We can simplify the product in parentheses using the fact that any number multiplied by its reciprocal equals one, to get

$$x \times 1 = 3.$$

The number 1 works as an identity for multiplication, meaning that the product of one with any other number results in the original number, regardless of whether 1 is on the right or the left. We have now arrived at our answer

$$x = 3$$

which tells us that the value of x must be three.

Now let's try to solve the additive version of this equation. We start with the equation

$$x + 2 = 6.$$

One approach might be to add negative two. This step works because we want to undo the addition of two, just like we undid the multiplication of two by canceling it out with one-half. In mathematics, one-half is called either the reciprocal or inverse of two, so we can think of -2 as an additive inverse of 2.

We arrive at the equation

$$(x + 2) + (-2) = 6 + (-2)$$

and by associativity on the left, and simply adding on the right, we get

$$x + (2 + (-2)) = 4.$$

Summing two with its additive inverse gives us

$$x + 0 = 4.$$

The sum of 0 with any number gives us that number, similarly to the way 1 behaves with multiplication. Thus we get

$$x = 4$$

and have found the value of x.

Let's summarize what we did in both cases. Though the equations and answers are different, the technique was exactly the same.

$x \times 2 = 6$	$x + 2 = 6$
EQUALITY	
$(x \times 2) \times \frac{1}{2} = 6 \times \frac{1}{2}$	$(x + 2) + (-2) = 6 + (-2)$
ASSOCIATIVITY	
$x \times (2 \times \frac{1}{2}) = 3$	$x + (2 + (-2)) = 4$
INVERSES	
$x \times 1 = 3$	$x + 0 = 4$
IDENTITY	
$x = 3$	$x = 4$

We can use this process whenever we have a collection of numbers with an associative operation that has both inverses and an identity. This will allow us to find x in the equation

$$x \star y = c$$

regardless of what operation the $\star$ symbol stands for.

In the process of solving these equations over the positive whole numbers, we used numbers such as one-half and negative two, which are clearly not positive whole numbers. Neither of those are a candidate for a number we can place in a cell, so in some sense this puzzle is not self contained in the set we're using. Similarly our clues, which are always positive whole numbers for additive and multiplicative puzzles over the set $\{1, 2, \cdots, n\}$, will not always be in the set itself. This is because sums and products of this set are generally larger than their individual parts, leading to further cases involving numbers outside of our solution set.

We might want to consider whether the set and operation we are using are really the best for creating and solving these types of puzzles. Perhaps there is a set of objects with an operation, which when applied always leads back the same set, so that the cell values and clues both come from the same collection. Imagine that this operation also satisfies the rules we used earlier for solving equations. Then there is always a solution to

$$x \star y = c,$$

for any y and c, so we can effectively solve for two-cell cages. Such a self contained mathematical structure, if it exists, would be the perfect place to construct and solve these types of puzzles.

The good, but by now predictable, news is that such structures do exist. These are what we call groups. Let's write down a formal definition of what it means to be a group. We can use the symbol $\star$ for our operation, regardless of whether it is usually written as "$\times$" or "$+$" or something else, and the symbol e for whatever our identity is, whether it be 1, 0, or its own thing.

A *group* is a set G together with a operation that we refer to here as $\star$. This operation maps each pair of elements a and b from G to some element $a \star b$ also in G. The operation must also satisfy three rules:

- *Associativity*: For any a, b, and c in G, we have $(a \star b) \star c = a \star (b \star c)$.
- *Identity*: There is some element in G which we call e, so that for every a in G, we have $a \star e = e \star a = a$.
- *Inverses*: For each element a in G, there is an inverse which we call a^{-1}, so that $a \star a^{-1} = a^{-1} \star a = e$.

Associativity is a statement about products of three elements, which can give meaning to expressions such as $a \star b \star c$, which might otherwise be ambiguous. However, this property says nothing about whether $a \star b$ equals $b \star a$ for pairs of elements a and b in the group. That property is known as *commutativity* and groups satisfying this property for any pair of elements are known as *abelian groups*.[1]

---

[1] This word comes from the name Niels Henrik Abel, who is one of multiple mathematicians who helped discover and form the field now called group theory. Along with Évariste Galois who discovered groups independently from Abel, he is considered one of the key creators of the subject. Both of these men are famous not just for their amazing mathematical accomplishments, but also for their tragic and interesting lives [8, 12, 17, 19].

Do groups actually exist? Just because we give a name to something that satisfies certain rules, doesn't mean something actually satisfies those rules. Fortunately, in this case, we don't need to worry. One learns about certain infinite groups early on in life. The first example we generally see is the set of whole numbers under addition, which forms a group with identity 0 where the inverse of each number x is $-x$. As this involves an infinite set, and the puzzles we wish to create are finite, this isn't the group we're looking for.

It turns out that there are plenty of finite groups. The number of elements in the set G is called the *order* of the group, and we will soon see that there are groups of order n for every positive whole number n. We will provide new examples of different types of finite groups in each of the sections ahead.

Abelian Groups

3

3.1 The Cyclic Groups $\mathbb{Z}_n$

Now that we know the rules that a finite group must follow, what might one look like? We start here by trying to make one over the set $\{0, 1, 2, 3, 4\}$ using addition. Due to our definition of group, we know our operation has to return values from this same set. For some values such as 1 and 3, we are okay, because $1 + 3 = 4$ which is on our set. For other values, such as 3 and 4, we're in a bit of trouble since $3 + 4$ equals 7, which is not in our set.

Possibly the most intuitive way around this, is to start over by wrapping around like a clock when our numbers get too large. For instance, we would replace the number 5 with the number 0, 6 with 1, 7 with 2, and so on, as illustrated by the table in Fig. 3.1. In such a system, a three cell cage containing the numbers 1, 3, and 4 would require a clue of 3.

With this operation of adding first and then wrapping around, the sum of elements in our set always stays within the set. However, does the mathematical structure we get meet the three criteria we set for being a group?

First, notice that just like in whole number addition, the identity here is zero. Adding zero to any number results in that number: no wrapping required. Our new operation is associative, since addition in the whole numbers is associative before anything gets wrapped. For inverses, we must show that for every element, there is an element that returns us to our identity of zero. Since

$$1 + 4 = 4 + 1 = 2 + 3 = 3 + 2 = 0 + 0 = 0$$

we can see that the numbers 1 and 4 are inverses of each other, the numbers 2 and 3 are inverses of each other, and the number 0 is its own inverse. This shows that the wrap-around system we've created does indeed form a group.

There's no reason we had to use the set $\{0, 1, 2, 3, 4\}$ in our construction. For any n bigger than 1, we can take the n numbers $\{0, 1, \cdots, n - 1\}$ and wrap around past

© The Author(s), under exclusive license to Springer Nature Switzerland AG 2024
D. Nacin, *The Group Theory Puzzle Book*, SUMS Readings,
https://doi.org/10.1007/978-3-031-70091-0_3

Old Number	0	1	2	3	4	5	6	7	8	9	10	11	12	13	14	15	$\cdots$
New Number	0	1	2	3	4	0	1	2	3	4	0	1	2	3	4	0	$\cdots$

Fig. 3.1 Table of values as we wrap around, setting every fifth number back to zero

that, setting n to 0, $n + 1$ to 1, and so on. This operation will still have zero as our identity, and be associative for the same reason as before. Finally, since the sum of k and $n - k$ is always n, which is equivalent to zero, the inverse of any nonzero k is simply $n - k$.

This structure we've defined here is often called *addition modulo n*, and referred to by the symbol $\mathbb{Z}_n$. The letter Z comes from "Zahl", the German words for number, and the small n reveals both the size of the group, as well as which number is now equivalent to 0 due to the wrap.[1] At times we may just wish to talk about this type of wrapping around with a different operation or no operation, and we will simply use the term *modulo n* for those instances.[2]

What's a quick way to find the correct element in $\mathbb{Z}_n$ when dealing with a large sum? For example, if we have a cage with cells containing the elements $1, 2, 3, 3, 4, 4$, resulting in a total sum of 17, what element would this be in our group $\mathbb{Z}_5$? As all multiplies of five are equivalent to zero, we know 15 is equivalent to zero. As 17 is two more than 15, it must be equivalent to 2. This whole process is equivalent to taking the remainder when dividing by 5 since we're looking for the amount that is left over after removing the largest possible multiple of 5. For any n, we can always divide by n and take the remainder to find the correct final value in $\mathbb{Z}_n$.

We'll actually be spending more time working in the other direction, since we'll be solving puzzles instead of creating them. Here, we're given a clue and have to figure out what possible numbers could fill the cells. For us, this means finding numbers with the same remainder. So for example, if our clue is 3 and we're in $\mathbb{Z}_5$ then the sum of the numbers in that cage could be 3, $3+5 = 8$, $3+10 = 13$, $3+15 = 18$, $\cdots$ and so on. Any collection of numbers summing to any of these values is a possible candidate for our cells.

Though it's not necessary for solving the puzzles ahead, let's briefly explain the name of this section. If all the elements in a finite group can be reached by applying the operation to some single element, then we call the group *cyclic*.[3] The groups

---

[1] The extra line in the letter Z comes from the font called Blackboard Bold. Bold was generally used on capital letters to denote famous sets of numbers. Since bold is hard to do on a blackboard in chalk, mathematicians resorted to adding extra lines to these characters on the chalkboard, which also later caught on in print [13].

[2] Computations modulo n may feel somewhat intuitive. If we can calculate what time it will be six hours after 8 o'clock, or how long a 98-minute movie is in hours and minutes, then we can already compute in groups such as $\mathbb{Z}_{12}$ and $\mathbb{Z}_{60}$.

[3] In the infinite case, the definition is slightly different. There, we say that a group is cyclic if every element can be reached using the operation from some element and its inverse, though in this book, we will avoid infinite puzzles.

$\mathbb{Z}_n$ are cyclic for each $n > 0$. To see this, start with the element 1 in our group and consider what happens when repeatedly apply our operation using only that element. We have $1, 1 + 1 = 2, 1 + 1 + 1 = 3$, and so on. Eventually we arrive at $n - 1$ and then wrap back to 0 after adding a total of n 1's together. These are all the elements of the group, and therefore our group is cyclic.[4]

If, and probably only if, one is having trouble with the next set of puzzles, then they might want to look at the questions below to get some ideas. We encourage the reader to jump right in first, before taking any of these into account.

1. For the group $\mathbb{Z}_4$, consider a cage with two cells. How many possibilities are there for the entries of those cells if the clue is odd? Is it the same when the clue is even, and if not, what are the values of the cells for each of the two even clues? What is the value of $0 + 1 + 2 + 3$ here, and how can this help us when combined with the fact that each row and column contains all four numbers? Is it sometimes possible to tell which cells contain even and odd numbers even when we do not know the specific values in those cells?
2. For the group $\mathbb{Z}_5$, consider a cage with two cells. For a given clue, is there always some number that cannot be in either of the cells? If so, is the number of forbidden entries the same for every clue? Which nearby values in the row or column that contains our cage might help restrict our possibilities even further? What is the value of $0 + 1 + 2 + 3 + 4$ here, and is there a way we can use that to help solve the puzzle?
3. For the group $\mathbb{Z}_6$, how many possibilities are there for a cage with two cells if the clue is even? What if it is odd? Instead of considering whether values come from the even set $\{0, 2, 4\}$ or the odd set $\{1, 3, 5\}$, are there other ways to group our six numbers that might help us in solving our puzzle? What is the value of $0 + 1 + 2 + 3 + 4 + 5$, and what role might that play in solving these puzzles?
4. For the groups $\mathbb{Z}_n$, with n greater than 6, how can we generalize the types of questions we asked for the previous cases?
5. The Latin square condition places constraints on each individual row and column. If some cages are contained entirely in that row or column, what information can we get about the values of the remaining cells in that row or column?
6. For anyone still stuck on a puzzle, there is a chapter near the end of this book for further tricks and techniques, where explanations behind certain methods are provided. Hopefully these tricks will arise naturally as readers develop strategies for exploring and interacting with these puzzles.

---

[4] Not only is $\mathbb{Z}_n$ always cyclic, it turns out that all finite cyclic groups are equivalent to $\mathbb{Z}_n$ for some value of n. However, what we mean by "equivalent" will not be explained until the next section.

Puzzle Rules

Fill in the cells so that each element of the cyclic group appears exactly once in every row and column. The clues in each cage indicate the sum in $\mathbb{Z}_n$ of the values from their cells. The following tables (Figs. 3.2, 3.3, 3.4, 3.5, 3.6, 3.7, and 3.8) may be helpful for solving these puzzles (Figs. 3.9, 3.10, 3.11, 3.12, 3.13, 3.14, 3.15, 3.16, 3.17, 3.18, 3.19, 3.20, 3.21, 3.22, 3.23, 3.24, 3.25, 3.26, 3.27, 3.28, 3.29, 3.30, and 3.31).

Fig. 3.2 The group operation table for the group $\mathbb{Z}_4$

+	0	1	2	3
0	0	1	2	3
1	1	2	3	0
2	2	3	0	1
3	3	0	1	2

Fig. 3.3 The group operation table for the group $\mathbb{Z}_5$

+	0	1	2	3	4
0	0	1	2	3	4
1	1	2	3	4	0
2	2	3	4	0	1
3	3	4	0	1	2
4	4	0	1	2	3

Fig. 3.4 The group operation table for the group $\mathbb{Z}_6$

+	0	1	2	3	4	5
0	0	1	2	3	4	5
1	1	2	3	4	5	0
2	2	3	4	5	0	1
3	3	4	5	0	1	2
4	4	5	0	1	2	3
5	5	0	1	2	3	4

Fig. 3.5 The group operation
table for the group $\mathbb{Z}_7$

+	0	1	2	3	4	5	6
0	0	1	2	3	4	5	6
1	1	2	3	4	5	6	0
2	2	3	4	5	6	0	1
3	3	4	5	6	0	1	2
4	4	5	6	0	1	2	3
5	5	6	0	1	2	3	4
6	6	0	1	2	3	4	5

Fig. 3.6 The group operation
table for the group $\mathbb{Z}_8$

+	0	1	2	3	4	5	6	7
0	0	1	2	3	4	5	6	7
1	1	2	3	4	5	6	7	0
2	2	3	4	5	6	7	0	1
3	3	4	5	6	7	0	1	2
4	4	5	6	7	0	1	2	3
5	5	6	7	0	1	2	3	4
6	6	7	0	1	2	3	4	5
7	7	0	1	2	3	4	5	6

Fig. 3.7 All possibilities for
two-cell cages in $\mathbb{Z}_8$

0	1	2	3	4	5	6	7
$1+7$	$0+1$	$0+2$	$0+3$	$0+4$	$0+5$	$0+6$	$0+7$
$2+6$	$2+7$	$3+7$	$1+2$	$1+3$	$1+4$	$1+5$	$1+6$
$3+5$	$3+6$	$4+6$	$7+4$	$5+7$	$2+3$	$2+4$	$2+5$
	$4+5$		$6+5$		$6+7$		$3+4$

Fig. 3.8 Inverses of
elements in $\mathbb{Z}_8$

a	0	1	2	3	4	5	6	7
a^{-1}	7	6	5	4	3	2	1	0

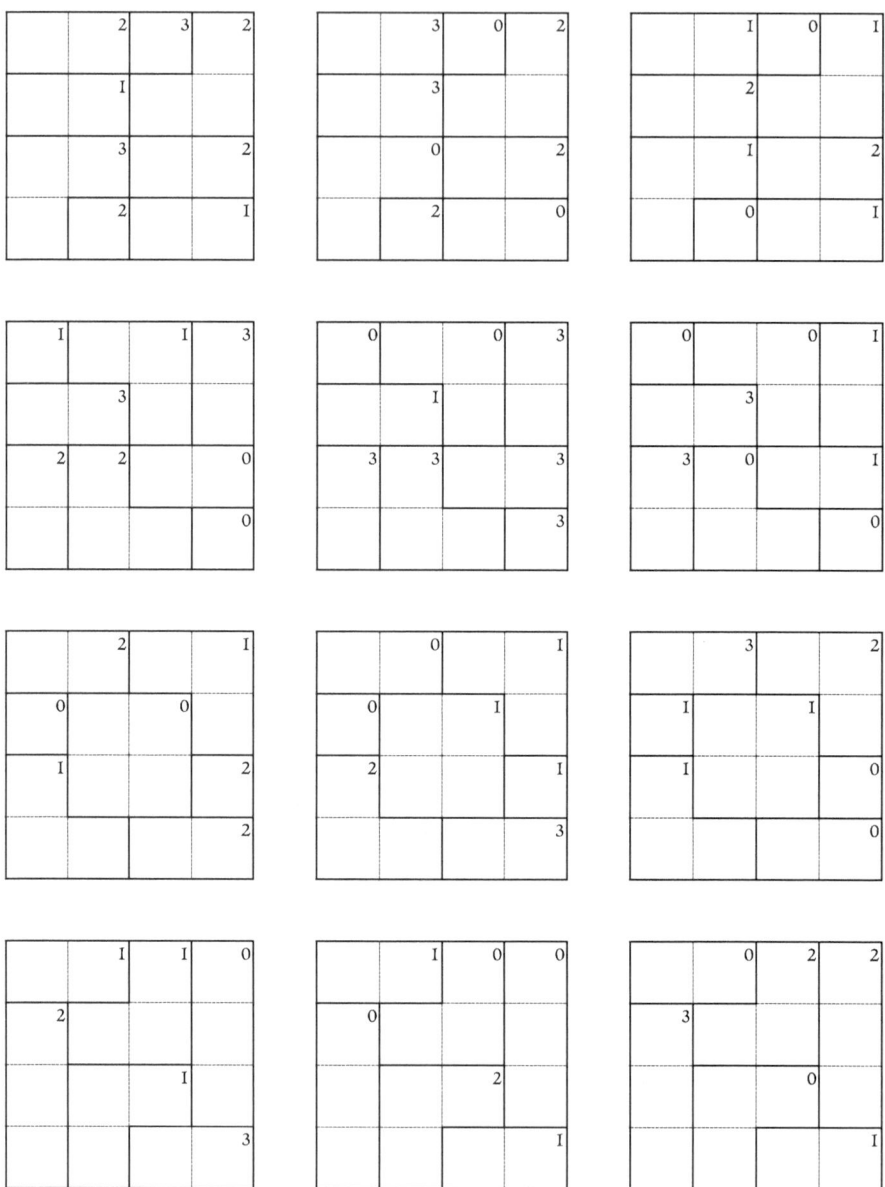

Fig. 3.9 Twelve puzzles over the cyclic group $\mathbb{Z}_4$

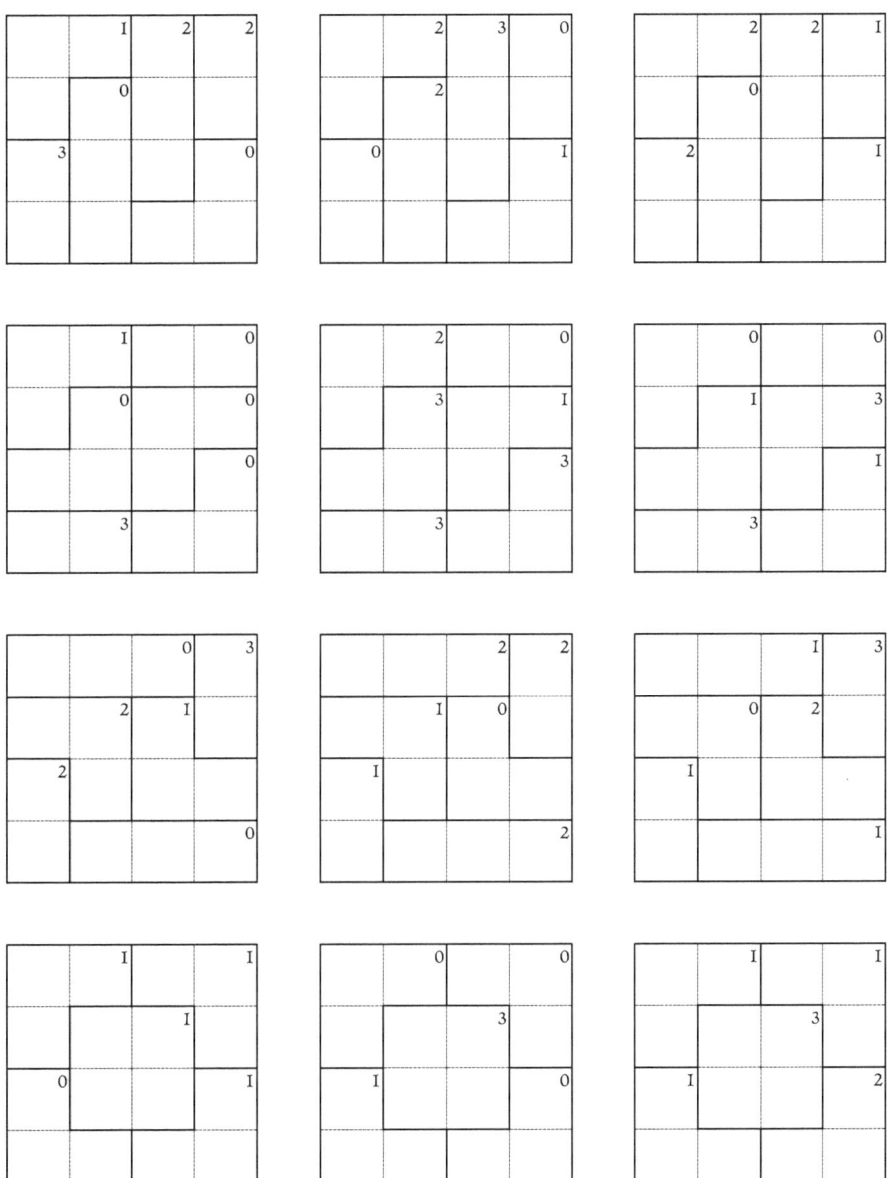

Fig. 3.10 Twelve puzzles over the cyclic group $\mathbb{Z}_4$

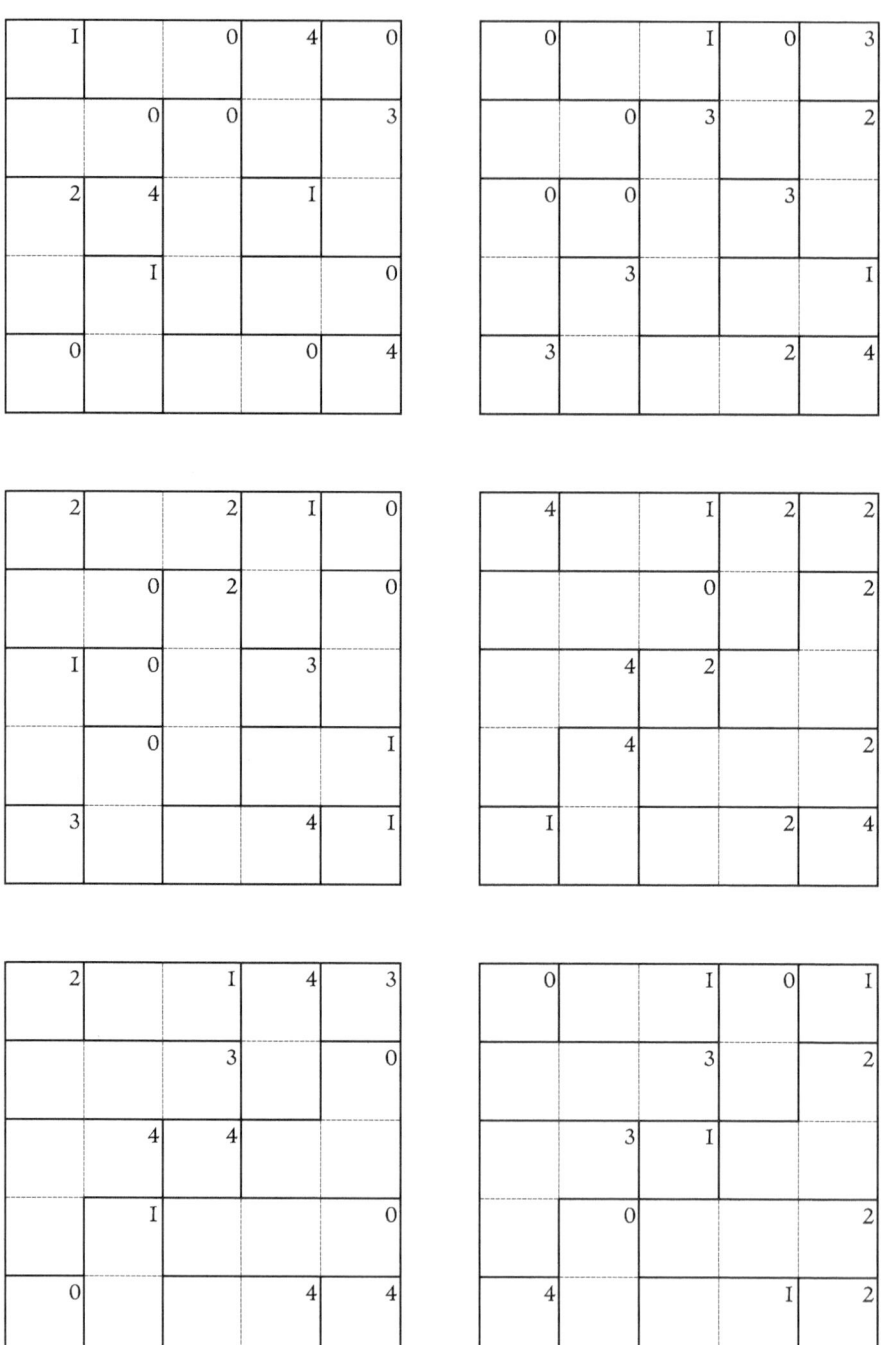

Fig. 3.11 Six puzzles over the cyclic group $\mathbb{Z}_5$

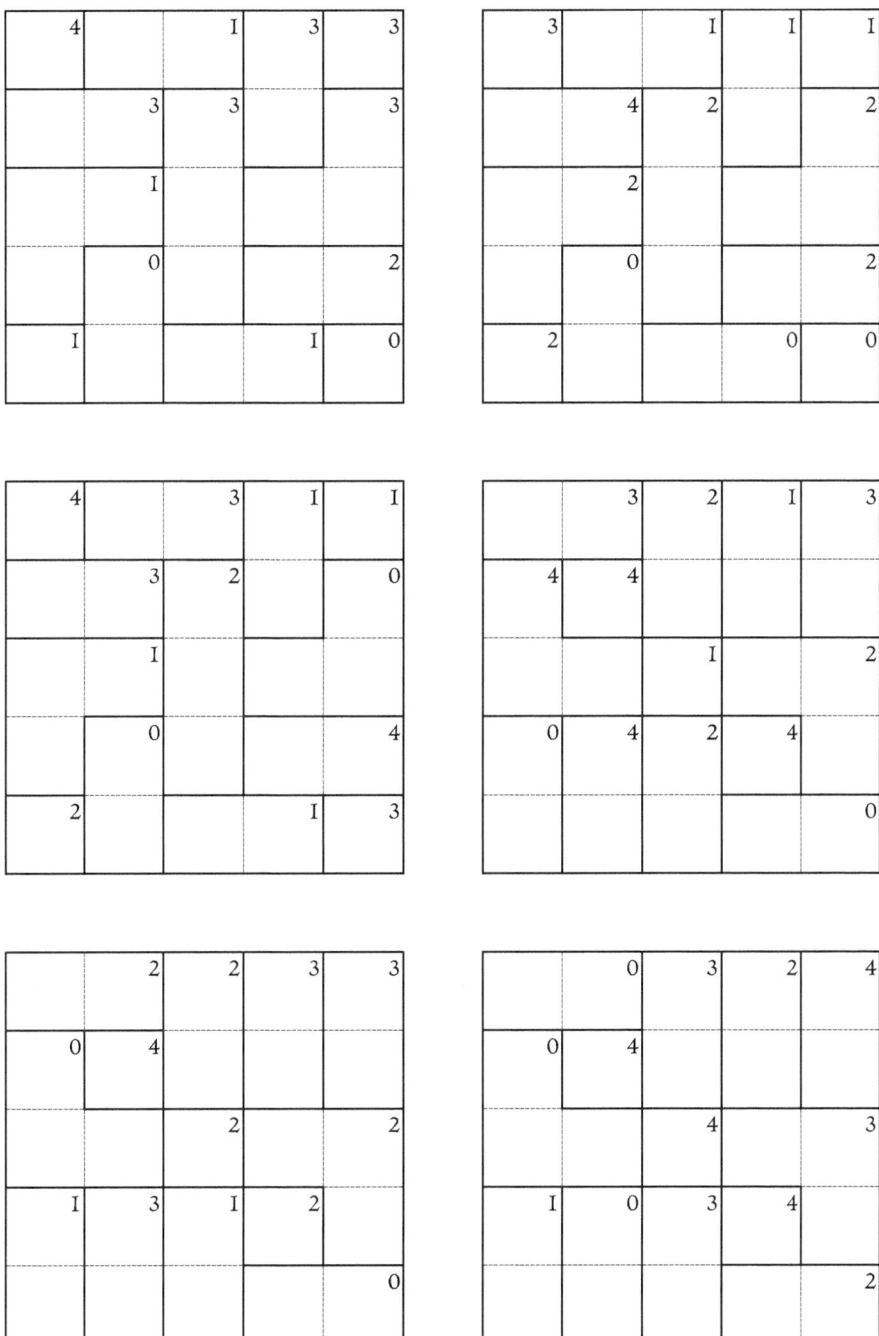

Fig. 3.12 Six puzzles over the cyclic group $\mathbb{Z}_5$

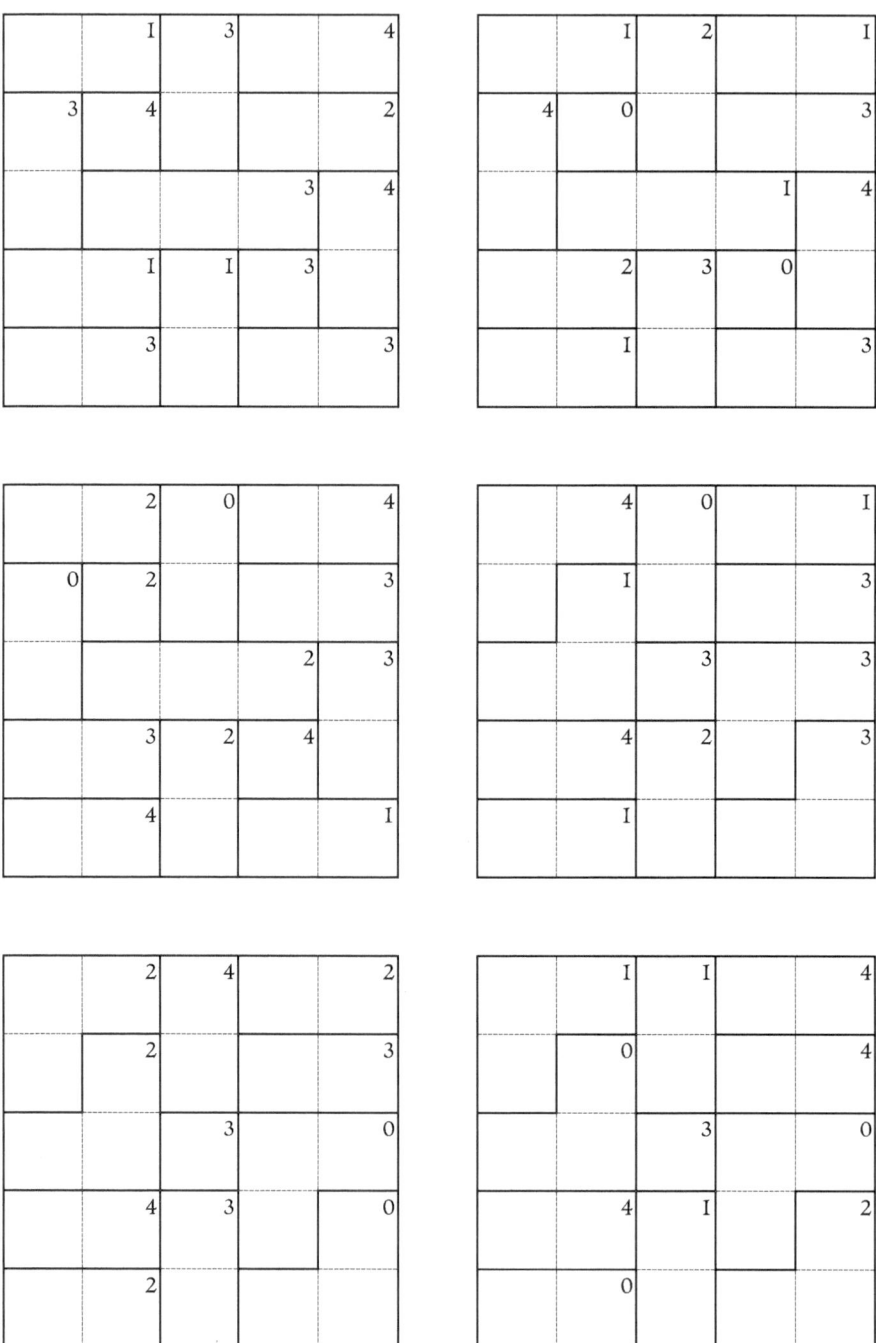

Fig. 3.13 Six puzzles over the cyclic group $\mathbb{Z}_5$

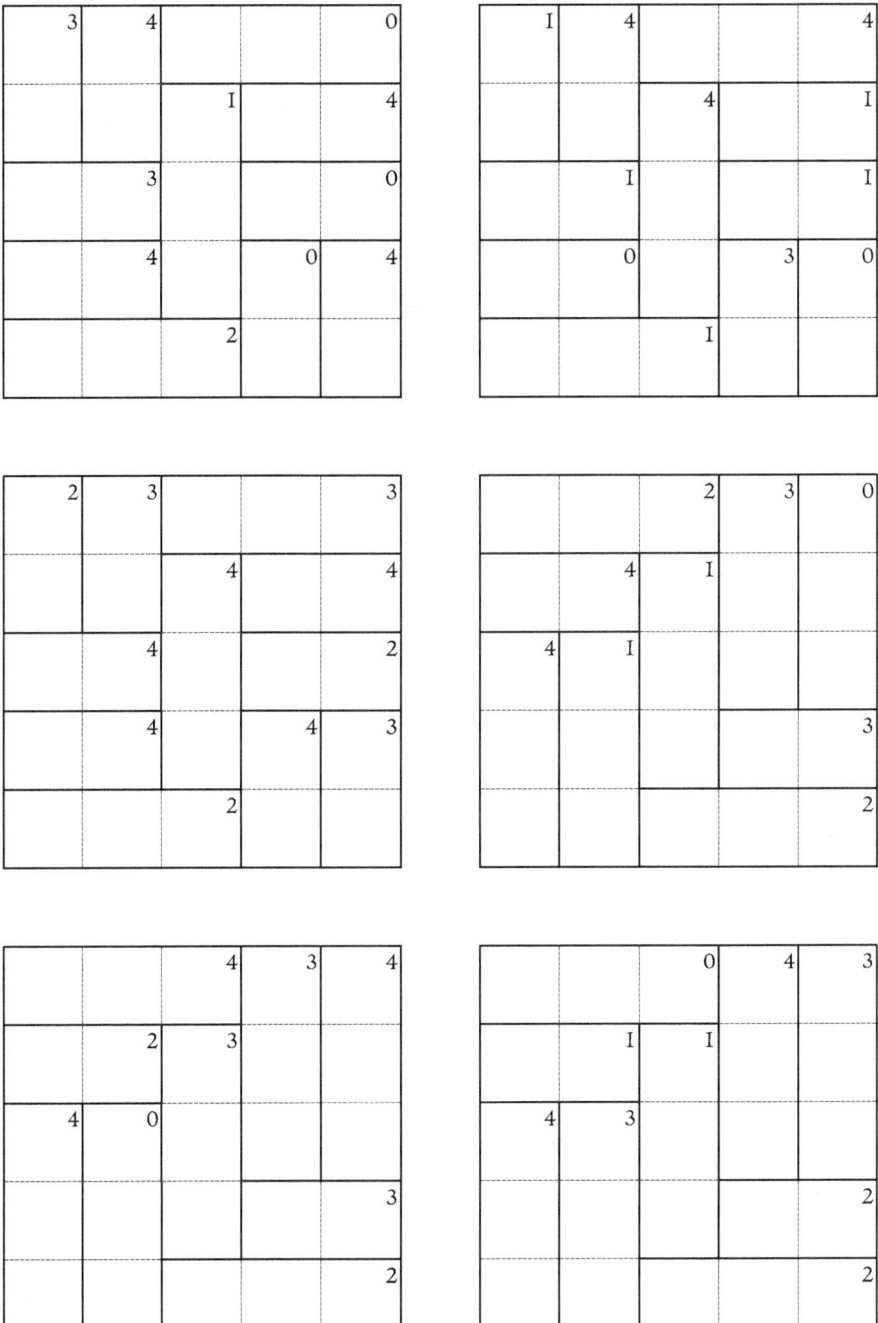

Fig. 3.14 Six puzzles over the cyclic group $\mathbb{Z}_5$

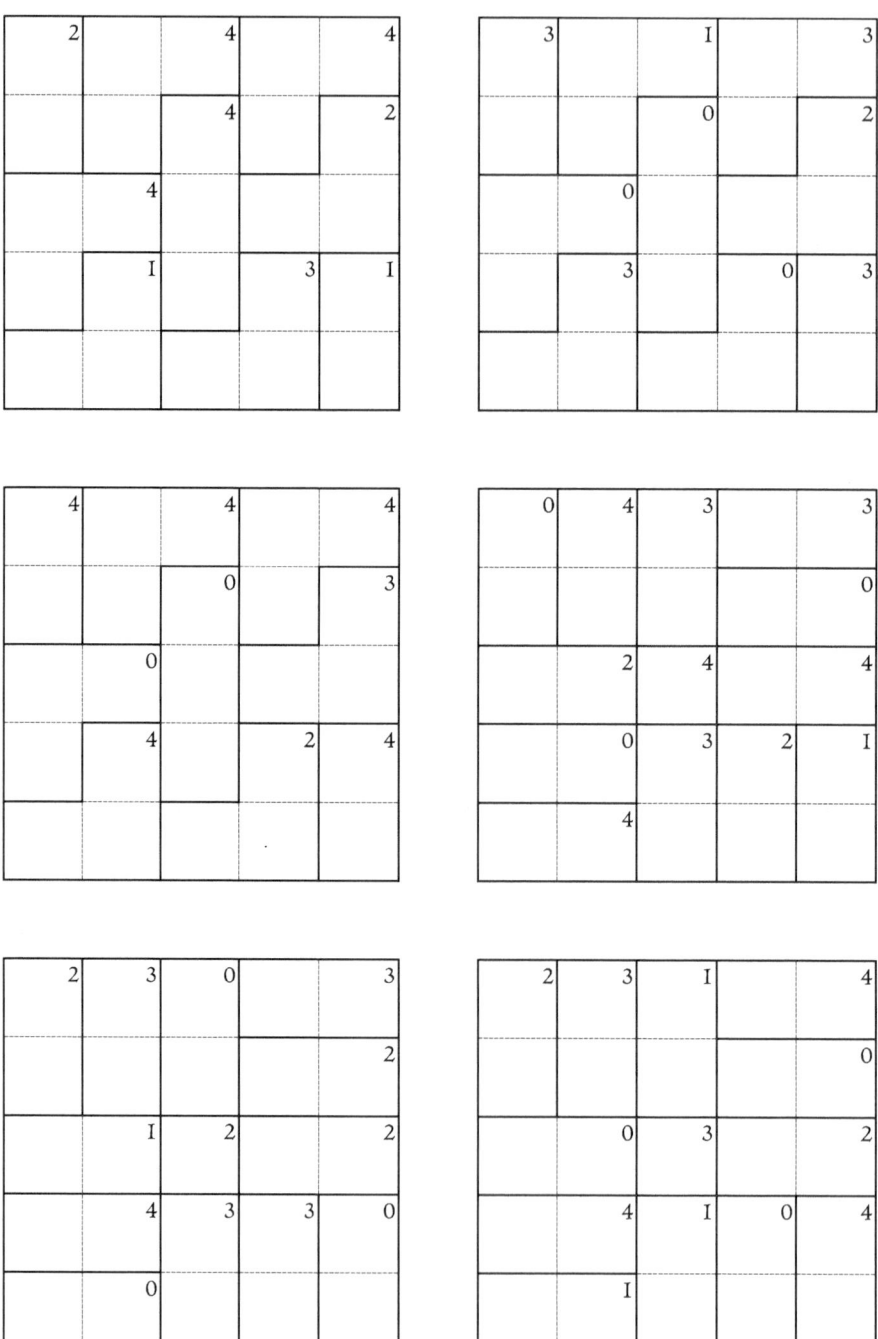

Fig. 3.15 Six puzzles over the cyclic group $\mathbb{Z}_5$

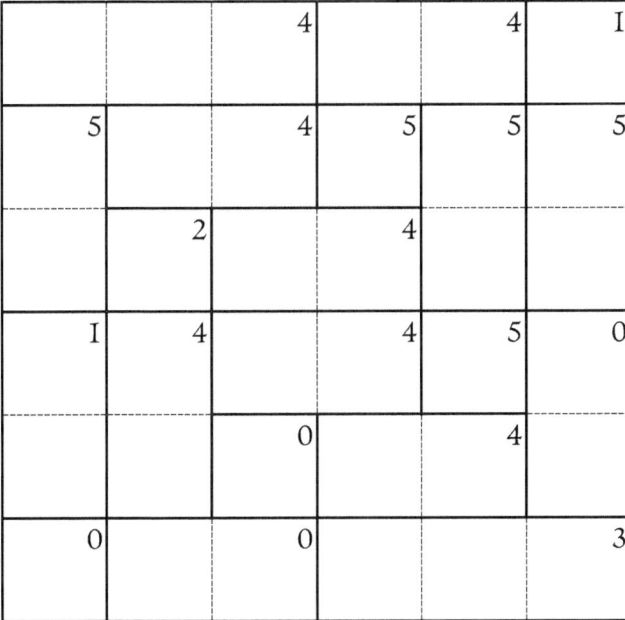

Fig. 3.16 Two puzzles over the cyclic group $\mathbb{Z}_6$

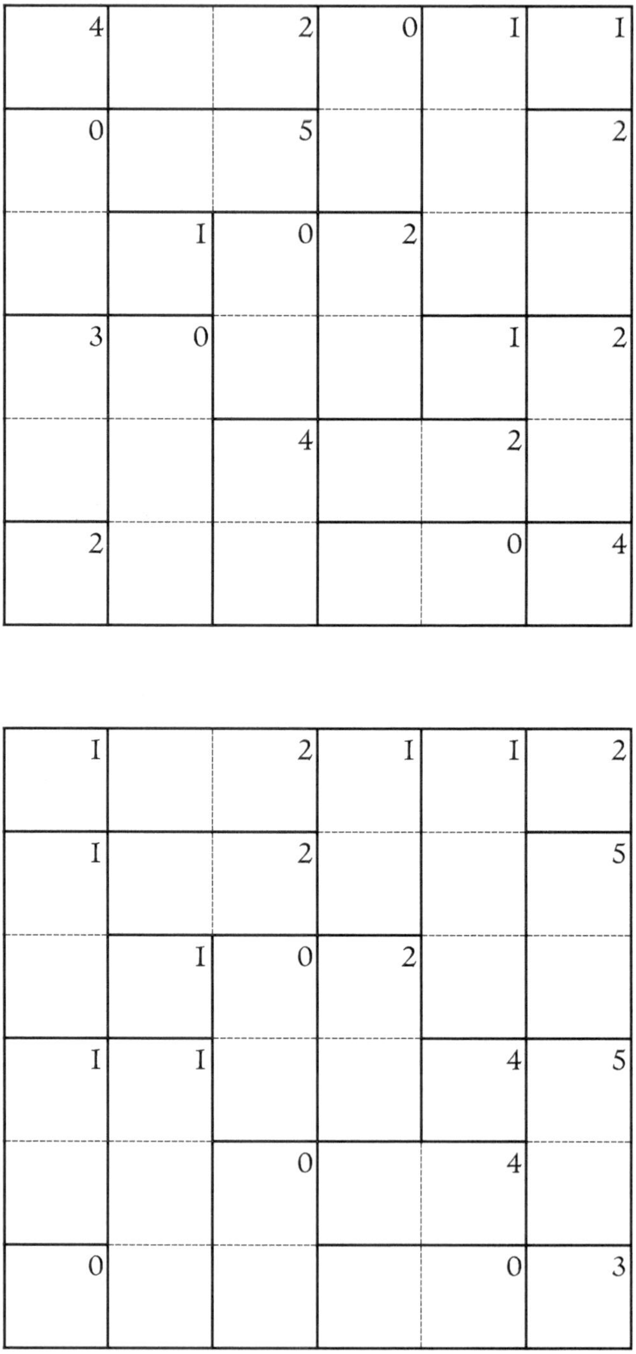

4		2	0	I	I
0		5			2
	I	0	2		
3	0			I	2
		4		2	
2				0	4

I		2	I	I	2
I		2			5
	I	0	2		
I	I			4	5
		0		4	
0				0	3

Fig. 3.17 Two puzzles over the cyclic group $\mathbb{Z}_6$

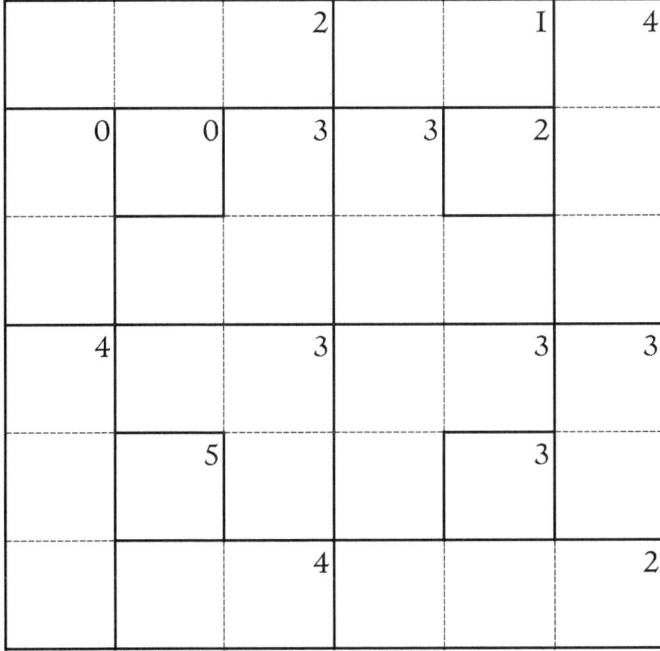

Fig. 3.18 Two puzzles over the cyclic group $\mathbb{Z}_6$

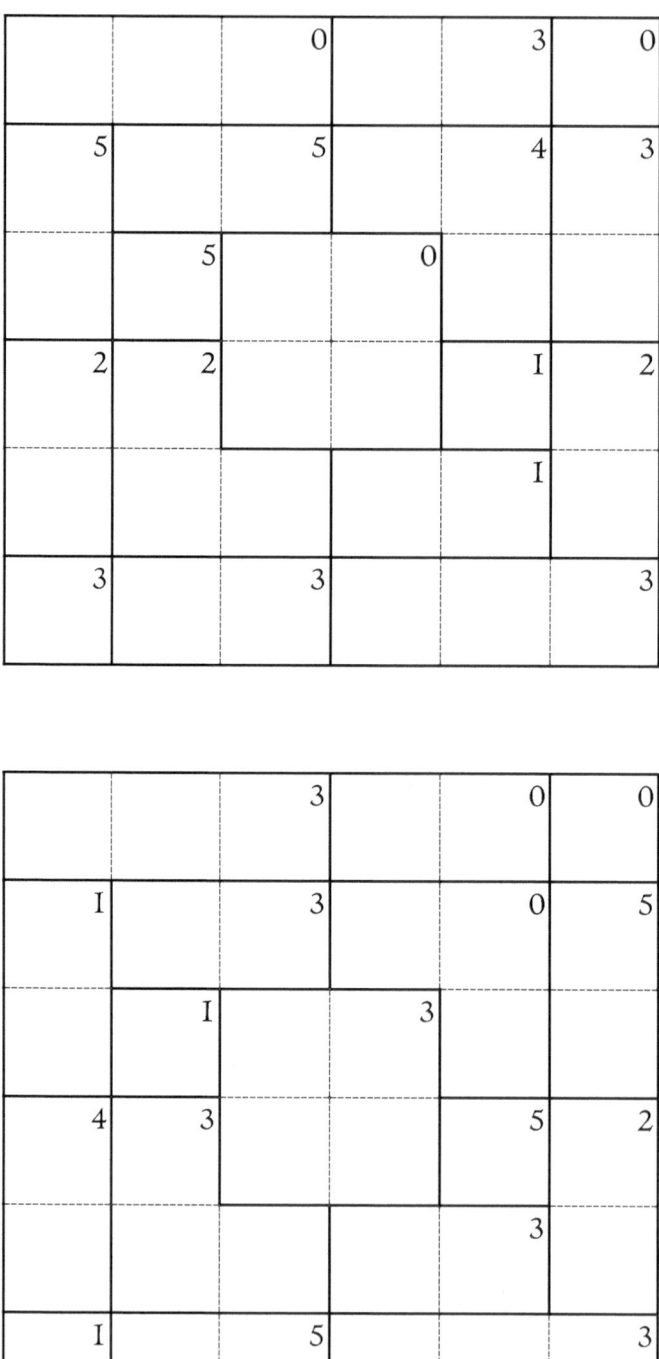

Fig. 3.19 Two puzzles over the cyclic group $\mathbb{Z}_6$

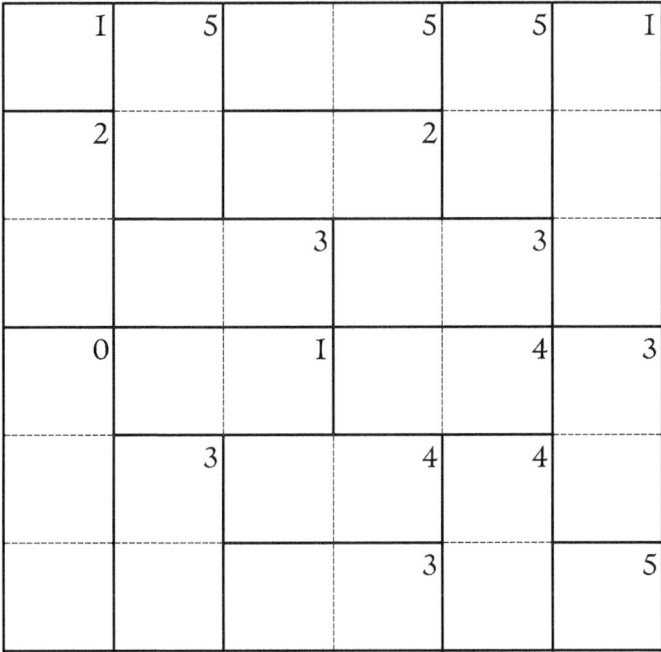

Fig. 3.20 Two puzzles over the cyclic group $\mathbb{Z}_6$

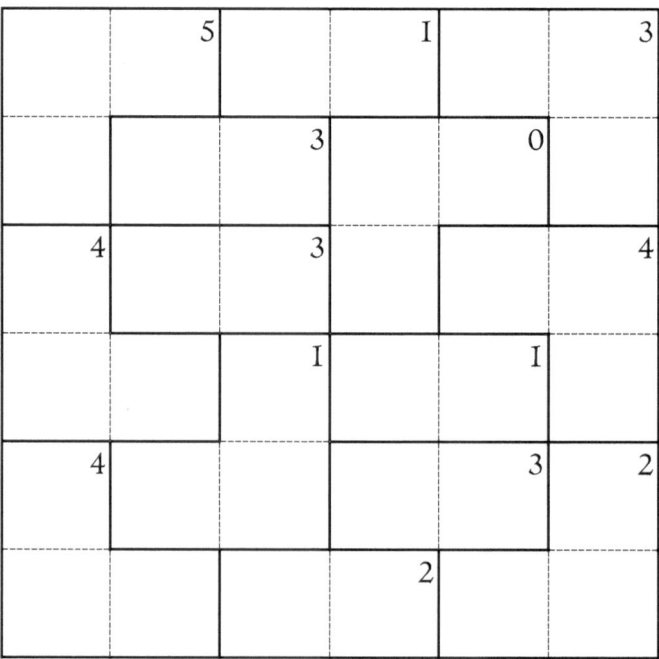

Fig. 3.21 Two puzzles over the cyclic group $\mathbb{Z}_6$

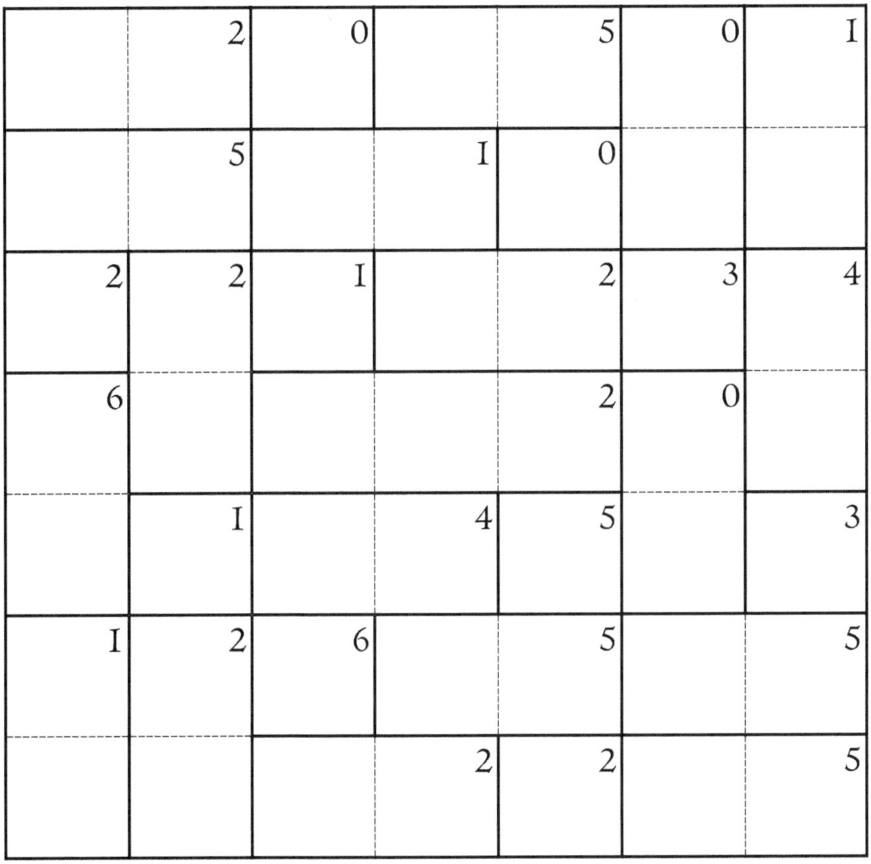

	2	0		5	0	I
	5		I	0		
2	2	I		2	3	4
6				2	0	
	I		4	5		3
I	2	6		5		5
			2	2		5

Fig. 3.22 A puzzle over the cyclic group $\mathbb{Z}_7$

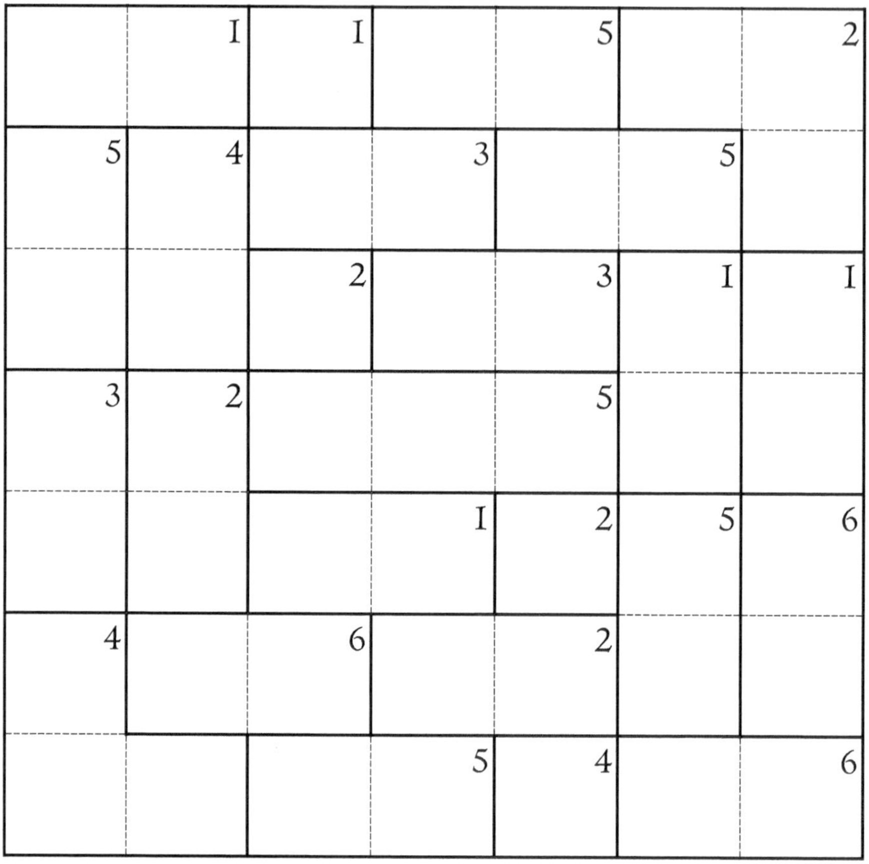

		I	I		5		2
	5	4		3		5	
			2		3	I	I
	3	2			5		
				I	2	5	6
	4		6		2		
				5	4		6

Fig. 3.23 A puzzle over the cyclic group $\mathbb{Z}_7$

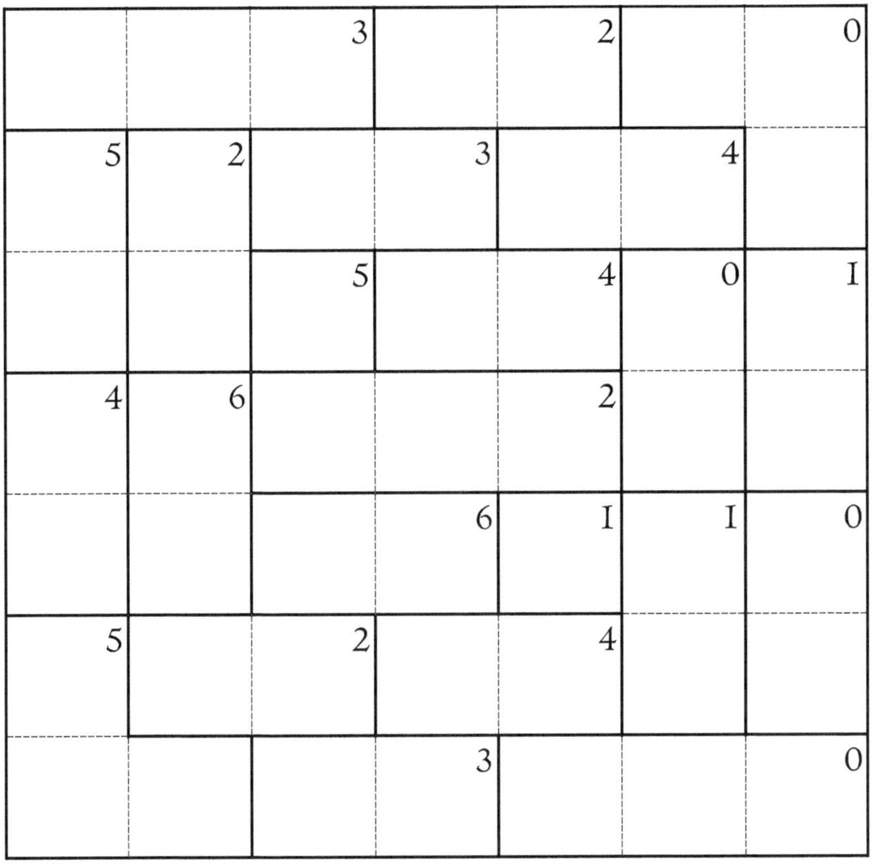

Fig. 3.24 A puzzle over the cyclic group $\mathbb{Z}_7$

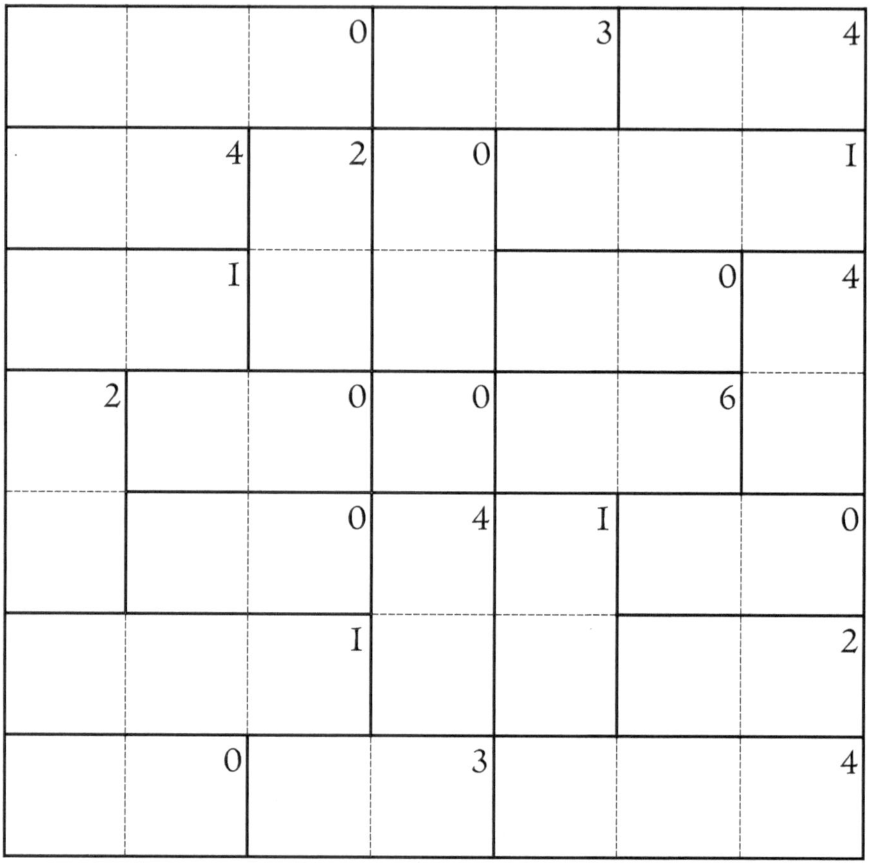

Fig. 3.25 A puzzle over the cyclic group $\mathbb{Z}_7$

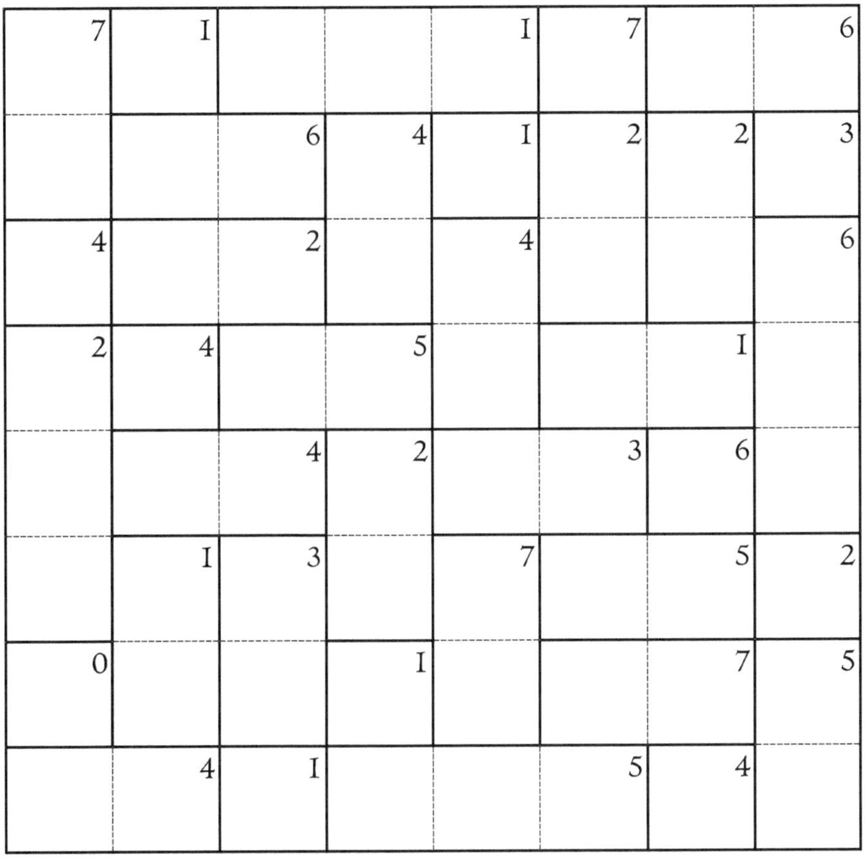

7	I			I	7		6
		6	4	I	2	2	3
4		2		4			6
2	4		5			I	
		4	2		3	6	
	I	3		7		5	2
0			I			7	5
	4	I			5	4	

Fig. 3.26 A puzzle over the cyclic group $\mathbb{Z}_8$

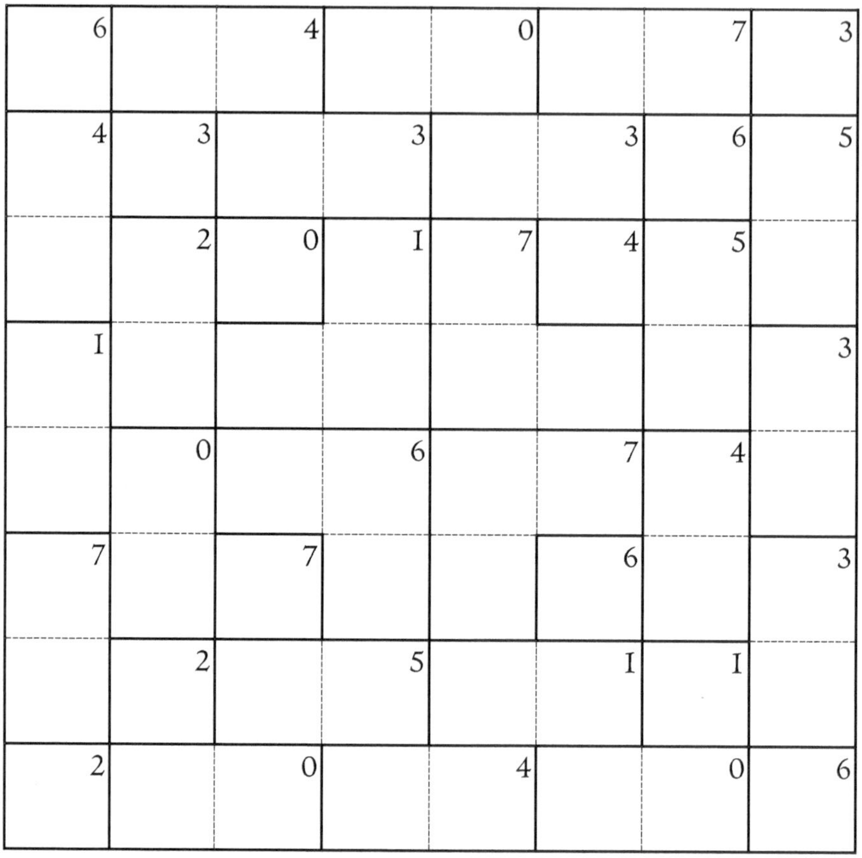

6		4		0		7	3
4	3		3		3	6	5
	2	0	I	7	4	5	
I							3
	0		6		7	4	
7		7			6		3
	2		5		I	I	
2		0		4		0	6

Fig. 3.27 A puzzle over the cyclic group $\mathbb{Z}_8$

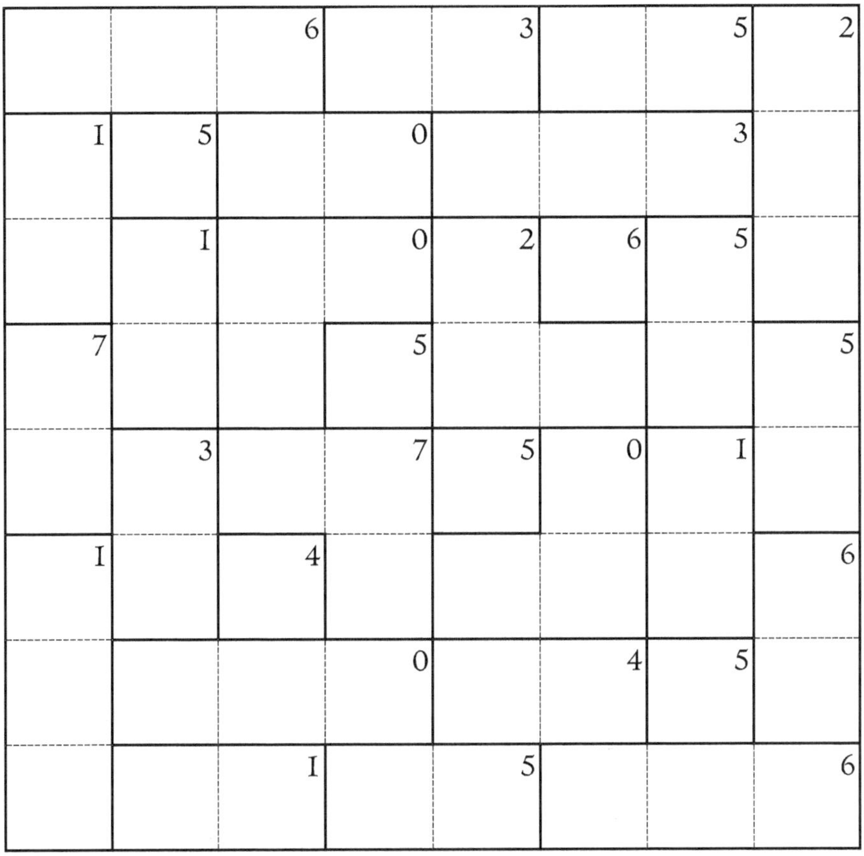

Fig. 3.28 A puzzle over the cyclic group $\mathbb{Z}_8$

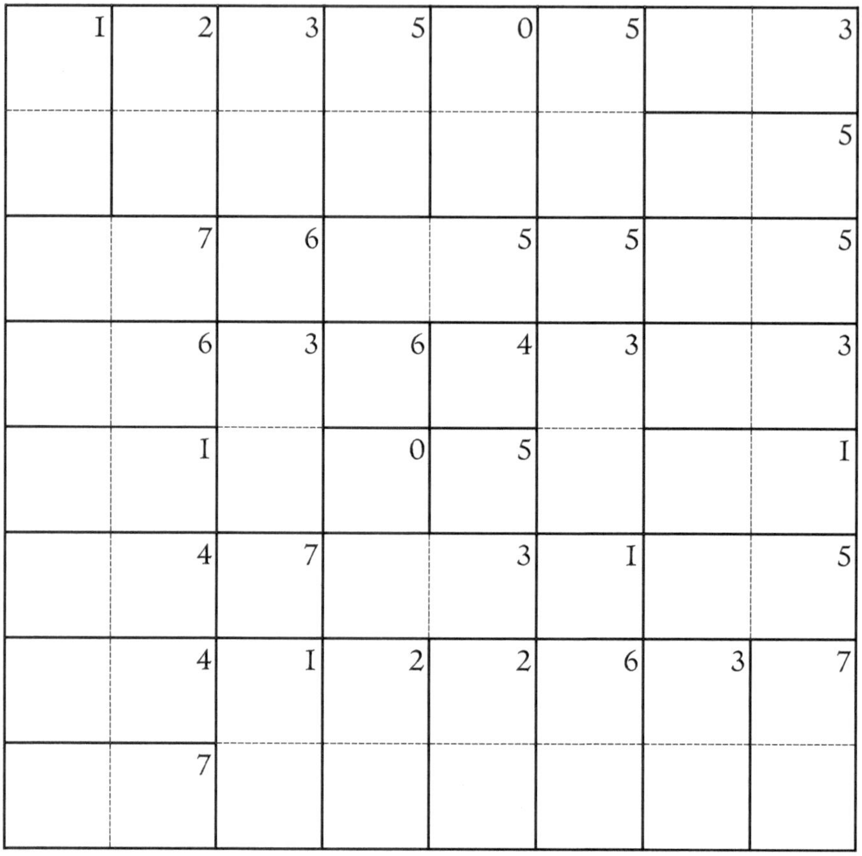

I	2	3	5	0	5		3
							5
	7	6		5	5		5
	6	3	6	4	3		3
	I		0	5			I
	4	7		3	I		5
	4	I	2	2	6	3	7
	7						

Fig. 3.29 A puzzle over the cyclic group $\mathbb{Z}_8$

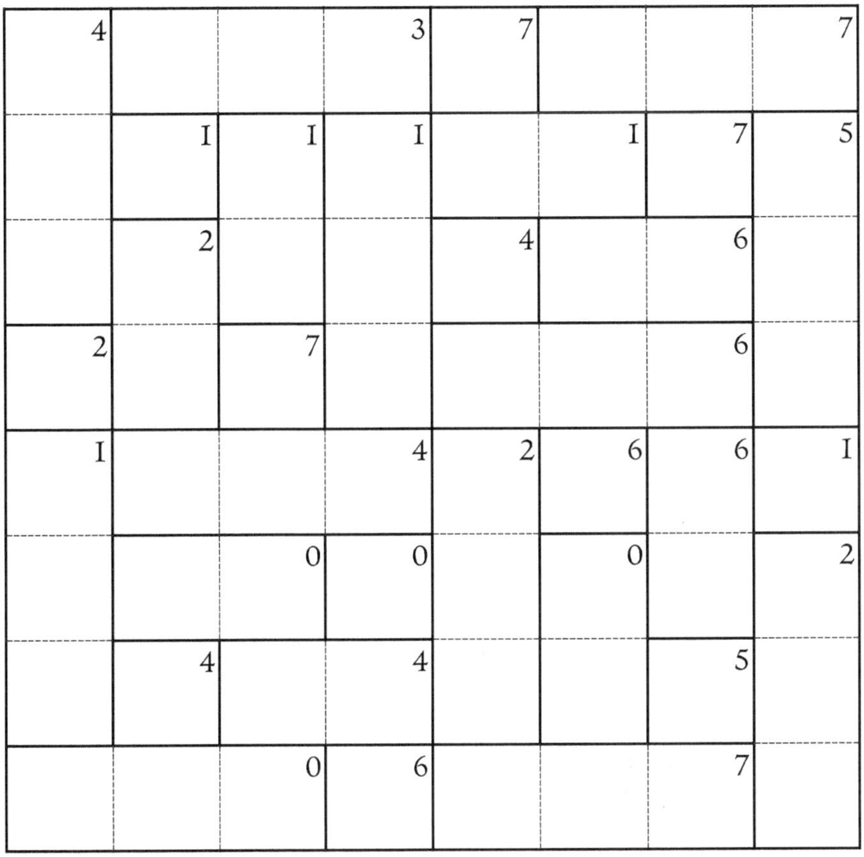

4				3	7			7
	1	1	1			1	7	5
	2			4			6	
2		7					6	
1			4	2	6		6	1
		0	0		0			2
	4		4			5		
		0	6				7	

Fig. 3.30 A puzzle over the cyclic group $\mathbb{Z}_8$

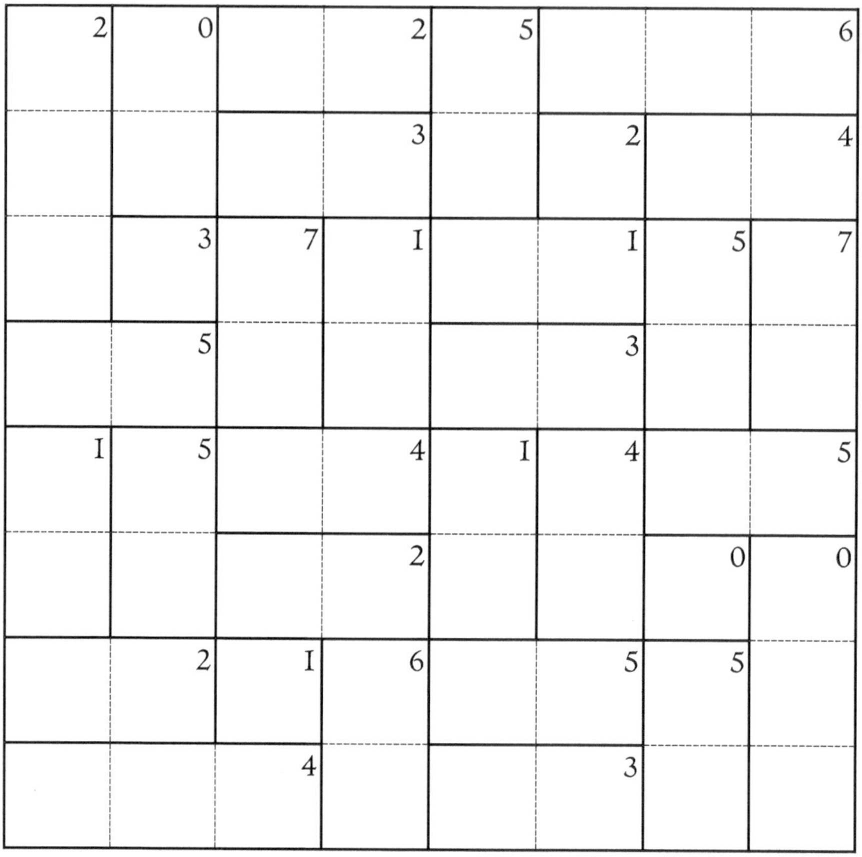

2	0		2	5			6
			3		2		4
	3	7	I		I	5	7
	5				3		
I	5		4	I	4		5
			2			0	0
	2	I	6		5	5	
		4			3		

Fig. 3.31 A puzzle over the cyclic group $\mathbb{Z}_8$

3.2   Other Cyclic Groups

We've seen that we can generate a rich source of puzzles through addition modulo n. Since multiplication is also fundamental, it is natural to ask what happens when we wrap around using that operation instead. Can we create solvable puzzles using multiplication modulo n?

The answer to this question is definitely yes, though some adjustments may be needed. If we take the set of numbers we use in $\mathbb{Z}_4$ and change the operation to multiplication, we get the table shown in Fig. 3.32. Notice that half of all products are equal to 0, so any clue of 0 gives little information. Meanwhile the clues 1 and 3 each only appear once as the product of an element with itself and thus will never appear as the clue for a two-cell cage. Though 1 acts as an identity, this structure is definitely not a group, as neither 0 nor 2 is invertible. We could still use this table to create puzzles with unique solutions if we construct them very carefully, but this might not be the best place to focus our attention.

Let's see if we fare better if we try multiplication modulo n with a different choice of n. Let n be any number bigger than 1. Regardless of what n is, the number 1 will act as our identity, since

$$1 \times a = a \times 1 = a$$

for any a in $\{0, 1, 2, \cdots, n-1\}$. This means the inverse of a is some number b so that $a \times b$ and $b \times a$ are both equal to 1. This is creates a problem when trying to find an inverse of 0, since

$$0 \times b = 0$$

for all values of b, and 0 is not equivalent to 1 modulo n.

Since 0 behaves badly here, what happens if we simply remove it? For example, instead of using $\{0, 1, 2, 3, 4\}$ modulo five, we can take the set $\{1, 2, 3, 4\}$ modulo five instead. Notice that we will never get an output of 0 by multiplying numbers from $\{1, 2, 3, 4\}$, since no product of those numbers will ever be a multiple of 5. If we make a table of products from this set, the values will also all be from this set.

Furthermore, since 1×1, 2×3, 3×2 and 4×4 all are equivalent to 1, all four elements are invertible. Multiplication is associative in the integers and thus will be

Fig. 3.32 Multiplication
over the set $\{0, 1, 2, 3\}$

×	0	1	2	3
0	0	0	0	0
1	0	1	2	3
2	0	2	0	2
3	0	3	2	1

Fig. 3.33 A multiplicative
group modulo five

×	1	2	3	4
1	1	2	3	4
2	2	4	1	3
3	3	1	4	2
4	4	3	2	1

Fig. 3.34 A very small
multiplicative group modulo
six

×	1	5
1	1	5
5	5	1

here as well. Now that we have removed the one element that lacks an inverse, what we have left is a group with the group operation table shown here in Fig. 3.33.

The trick of removing zero does not always result in a group. In our attempt with $\mathbb{Z}_4$ the element 2 also lacked an inverse. Even though we must always remove 0, this isn't usually the only number modulo n without an inverse.

Consider multiplication modulo 6, where an inverse of 2 would be some number such that $2 \times k$ is equivalent to 1. Since $2 \times k$ is always even, and all the numbers equivalent to 1 modulo 6 are odd, this is impossible. The number 4 cannot have an inverse either for a similar reason. An inverse of 3 would be some number where $3 \times k$ is equivalent to 1, which is impossible since the positive numbers equivalent to 1 are in the set

$$\{1, 7, 13, 19, 25, \cdots \}$$

which contains no multiples of 3. This leaves us with only the numbers 1 and 5, each of which is their own inverse modulo 6. We are left with the disappointing group shown in Fig. 3.34.

Why were we able to keep four of our numbers modulo 5 but only two of our numbers modulo 6? Is there any simple pattern to determine which numbers will and won't have inverses ahead of time? Let's try looking at one more example, and try to form a multiplicative group modulo 9.

Working modulo nine, we have to remove 0 as usual, but what about the numbers 3 and 6? Any multiple of these will be divisible by 3. The positive numbers equivalent to 1 are

$$\{1, 10, 19, 28, 37, 46, \cdots \},$$

Fig. 3.35 A comparison of $U(5)$ and $\mathbb{Z}_4$

×	1	2	3	4
1	1	2	3	4
2	2	4	1	3
3	3	1	4	2
4	4	3	2	1

+	0	1	2	3
0	0	1	2	3
1	1	2	3	0
2	2	3	0	1
3	3	0	1	2

a set which contains no multiples of 3. The rest of the numbers seem okay though, as 2 and 5 are inverses of each other, 4 and 7 are inverses of each other, and 1 and 8 are inverses of themselves. What went wrong with 3 and 6 here, and could it have something to do with the fact that those numbers both share the divisor 3 with 9? After all, all the numbers we removed modulo 6 were divisible by either 2 or 3, which are divisors of 6.

How can we tell which elements of $\{0, 1, 2, \cdots, n - 1\}$ will be invertible when constructing a multiplicative group modulo n? The answer is to throw away any number that shares a divisor with n that is bigger than one.[5]

For example, if we wanted to construct a multiplicative group modulo 10, we could start by finding the divisors of 10 which are greater than one. These are 2, 5, and 10, so we would need to remove all the numbers divisible by any of those. This means removing $\{0, 2, 4, 5, 6, 8\}$ and leaves us with the set $\{1, 3, 7, 9\}$. Notice that zero will always get removed in this process because even though no positive integer is divisible by zero, every positive integer divides zero.

We can perform this construction for any positive integer n. The set of remaining numbers under the operation of multiplication modulo n is what we will call the *group of units*, $U(n)$.[6] We now have a new class of groups, so we should be able to create exciting new puzzles, except we might first want to examine how new some of these groups actually are.

Let us compare our table for $U(5)$ next to our table for $\mathbb{Z}_4$, both shown in Fig. 3.35.

These both look different enough. They share some of the same symbols, but they each have distinctive characteristics apart from the symbols being used. The first table has only two distinct entries in the center four squares, whereas the second has three. The second table has the diagonal entries, from bottom left to top right, all equal to the same number, whereas the first does not. These seem to be very different groups, but what happens when we reorganize things a bit?

---

[5] The beautiful proof that this works is contained in a standard first course in either abstract algebra or number theory and involves greatest common divisors [6, 7, 16].

[6] In certain mathematical structures where some elements are multiplicatively invertible and others are not, we call the invertible elements the *units*. To construct these multiplicative groups modulo n, we had to disregard the non-units, and thus the name is appropriate.

Fig. 3.36 A second comparison of $U(5)$ and $\mathbb{Z}_4$

×	1	2	4	3
1	1	2	4	3
2	2	4	3	1
4	4	3	1	2
3	3	1	2	4

+	0	1	2	3
0	0	1	2	3
1	1	2	3	0
2	2	3	0	1
3	3	0	1	2

Fig. 3.37 A relabeling of the elements of $U(5)$

$$U(5) \quad \mathbb{Z}_4$$
$$1 \rightarrow 0$$
$$2 \rightarrow 1$$
$$3 \rightarrow 3$$
$$4 \rightarrow 2$$

There is no reason we needed to write these numbers in any particular order in either of the tables. In the $\mathbb{Z}_4$ table, we started with the identity in the first row and column. We next took 1, followed by $1 + 1 = 2$, and then $1 + 1 + 1 = 3$. The ordering we chose goes along with applying the group operation iteratively to a single element. What if we used a similar approach for ordering $U(5)$ on the left? We can again start with the identity, then proceed with 2, followed by $2 \times 2 = 4$, and then $2 \times 2 \times 2$. We would get the tables shown in Fig. 3.36.

Now let's examine both of these tables closely. They are not identical, because they use different symbols, but observe the positions of these symbols on each table. If we transform all the 1's into 0's, all the 2's into 1's, and all the 4's into 2's, then we could convert the first table into the next one exactly! Even though the groups first appeared different, the only differences were the symbols that we used and their order. The underlying group structure here was the same.

We can illustrate the map between elements that transforms $U(5)$ into $\mathbb{Z}_4$ with a diagram, as shown in Fig. 3.37. Notice that if we reverse the direction of the arrows, we get a set of instructions for turning $\mathbb{Z}_4$ into the group $U(5)$. It makes sense that if one group is equivalent to a second group after relabeling, then the second group must also be equivalent to the first in the same way.

Suppose we have a puzzle over $U(5)$ and we would rather solve a puzzle over $\mathbb{Z}_4$. We could always choose to convert the clues from those in $U(5)$ over to clues in $\mathbb{Z}_4$, solve the puzzle in $\mathbb{Z}_4$, and then convert the solution back. The mathematical difficulty of the puzzle will be equivalent over either group, although personal preference may lead one to feel more comfortable working in one group over the other.

Fig. 3.38 A group G
equivalent to $U(5)$, and hence
$\mathbb{Z}_4$

+	0	2	4	6
0	0	2	4	6
2	2	4	6	0
4	4	6	0	2
6	6	0	2	4

G	$U(5)$	$\mathbb{Z}_4$
$0 \to$	$1 \to$	0
$2 \to$	$2 \to$	1
$6 \to$	$3 \to$	3
$4 \to$	$4 \to$	2

Fig. 3.39 The group $U(10)$
and the map showing it is
isomorphic to $\mathbb{Z}_4$

×	1	3	9	7
1	1	3	9	7
3	3	9	7	1
9	9	7	1	3
7	7	1	3	9

$U(10)$	$\mathbb{Z}_4$
$1 \to$	0
$3 \to$	1
$9 \to$	2
$7 \to$	3

Suppose we have yet another group G which can be transformed into $U(5)$.[7] One such group can be formed from the even numbers modulo 8 under addition. Because $U(5)$ is transformable into $\mathbb{Z}_4$, G will also be transformable into $\mathbb{Z}_4$. To find the direct labeling from G to $\mathbb{Z}_4$, we just have to combine the two relabelings together as shown in Fig. 3.38. Because we can combine relabelings, we can start building up collections of groups all of which are structurally equivalent. If we can show a new group is equivalent to just one group from a collection of equivalent groups, we then know the new group is equivalent to all others in the collection.

When we have two groups G and H that are the same when elements are relabeled, we call them *isomorphic*.[8] A map which relabels the elements of two isomorphic groups is called an *isomorphism*.

Another group isomorphic to $\mathbb{Z}_4$ is the group $U(10)$. To form this group of units, we remove numbers that share divisors with 10 leaving us only with 1,3,7 and 9. Figure 3.39 shows the table for $U(10)$ together with the isomorphism onto $\mathbb{Z}_4$.

---

[7] G is a good disposable name for any group which just needs to show up for a single example, which is what we have here.

[8] Using the notation of functions, we can say two groups are isomorphic if there is a bijection f from G to H so that

$$f(xy) = f(x)f(y)$$

for any elements x and y in G. This states that if we multiply in G and then map the result over to H, we get the same thing as if we mapped both elements over first and then performed the multiplication in H. We can think of f as the map that relabels the elements of G.

Since these groups are essentially equivalent to $\mathbb{Z}_4$, they are all still cyclic. We can reach all of the elements by applying the operation to some single element. For the evens modulo eight, both 2 and 6 allow us to reach the whole group, whereas for $U(10)$, both 3 and 7 do the trick. For example, starting with 7 yields $7, 7 \times 7 = 9, 7 \times 7 \times 7 = 3,$ and $7 \times 7 \times 7 \times 7 = 1$ resulting in $\{7, 9, 3, 1\} = \{1, 3, 7, 9\}$, which is the entire group.

The group $\mathbb{Z}_4$ is not an exception. There are numerous instances where seemingly different groups turn out to be equivalent to $\mathbb{Z}_n$ for some n. For example, the groups $U(7)$, $U(9)$, $U(14)$ and $U(18)$ are all equivalent to $\mathbb{Z}_6$. We provide a handful of puzzles over $U(5)$, $U(7)$, $U(9)$ and $U(10)$ at the end of this section, despite the fact that they are equivalent to groups from the previous section. If the reader wishes, they could convert these puzzles back over to $\mathbb{Z}_n$ by changing each of the clues to the appropriate element. We encourage readers to attempt these puzzles as they are, noticing the similarities to puzzles with addition modulo n, and considering how techniques discovered for that setting can still be applied here.

At this point in this text, we are now discussing some groups that use '+' as the operation symbol and others that use '×'. We would like to avoid having to say "either the sum or product or possibly something else" each time we refer to the general operation of some unknown group. Because of this, we refer to an arbitrary group operation as our product regardless of the actual operation. We may also use product notation for any operation. For example, a^{-1} will always be the inverse of the element a even if a is in $\mathbb{Z}_n$, and thus the actual inverse of a is $-a$. The element a^2 represents the result of applying the group operation on a with itself, regardless of whether that means $a + a$ or $a \times a$, depending on the operation of the group. This may take some time getting used to, but it goes along with the notation for our original group definition, and is a small sacrifice for being able to better talk about groups in generality.

One stated goal of this text is to introduce puzzles over every group of order eight or less. What exactly do we mean by that? After all, if we have one group over the set $\{e, a, b, c\}$, we could make a new group by renaming the elements something like $\{w, x, y, z\}$ or even {Eve, Anna, Bob, Carla}. Because of this, one could claim there are infinitely many groups of order eight or less. What we actually mean is that we will introduce puzzles over every structural possibility for a group. That is, any group of order eight or less will be isomorphic to some puzzle from this book. Are all groups with the same number of elements isomorphic to each other? We will answer this question in the next section.

Puzzle Rules

Fill in the cells so that each element of the cyclic group appears exactly once in every row and column. The clues in each cage indicate the product of the values from their cells. The following tables (Figs. 3.40, 3.41, and 3.42) may be helpful for solving these puzzles (Figs. 3.43, 3.44, 3.45, and 3.46).

Fig. 3.40 The group operation tables for the cyclic groups $U(5)$ and $U(10)$

×	1	2	3	4
1	1	2	3	4
2	2	4	1	3
3	3	1	4	2
4	4	3	2	1

×	1	3	7	9
1	1	3	7	9
3	3	9	1	7
7	7	1	9	3
9	9	7	3	1

Fig. 3.41 The group operation table for the cyclic group $U(7)$

×	1	2	3	4	5	6
1	1	2	3	4	5	6
2	2	4	6	1	3	5
3	3	6	2	5	1	4
4	4	1	5	2	6	3
5	5	3	1	6	4	2
6	6	5	4	3	2	1

Fig. 3.42 The group operation table for the cyclic group $U(9)$

×	1	2	4	5	7	8
1	1	2	4	5	7	8
2	2	4	8	1	5	7
4	3	8	7	2	1	4
5	4	1	2	7	8	5
7	5	5	1	8	4	2
8	6	7	4	5	2	1

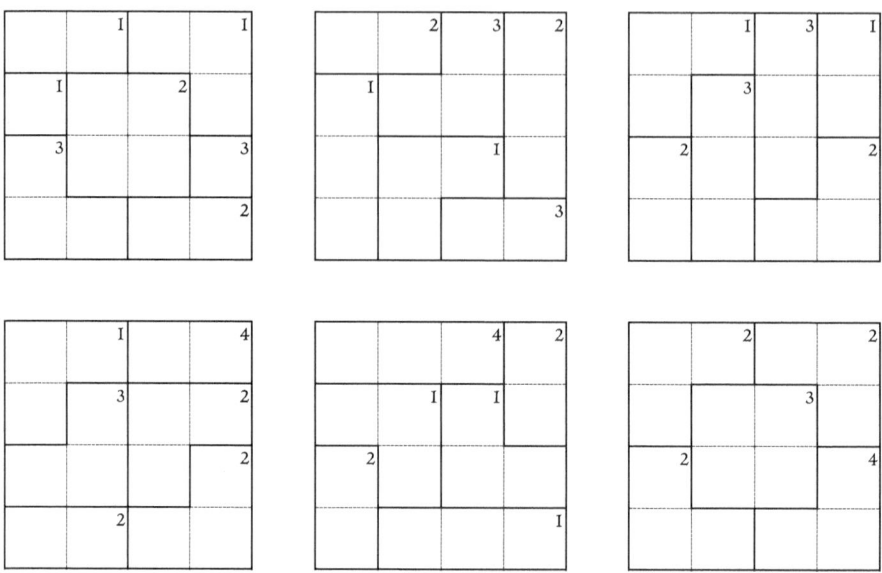

Fig. 3.43 Six puzzles over the group $U(5)$

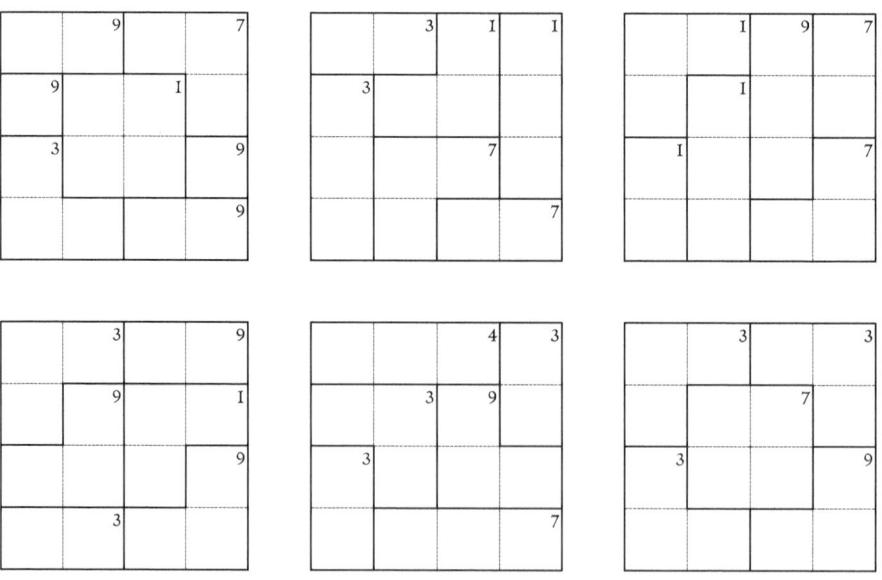

Fig. 3.44 Six puzzles over the group $U(10)$

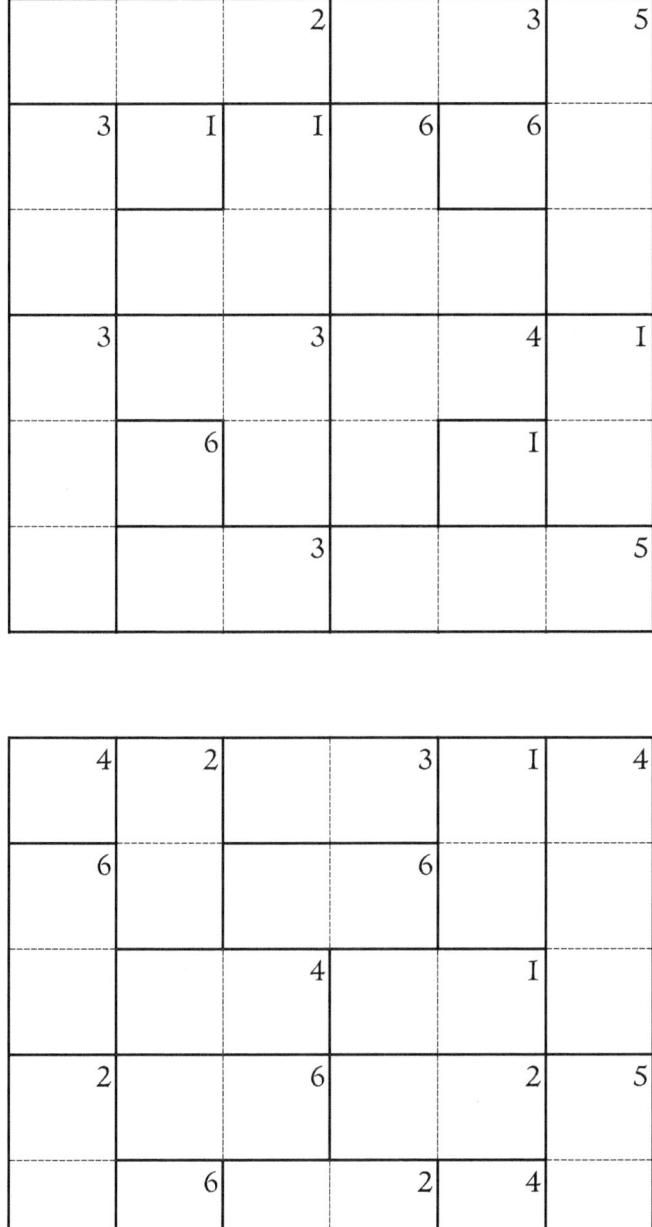

Fig. 3.45 Two puzzles over the group $U(7)$

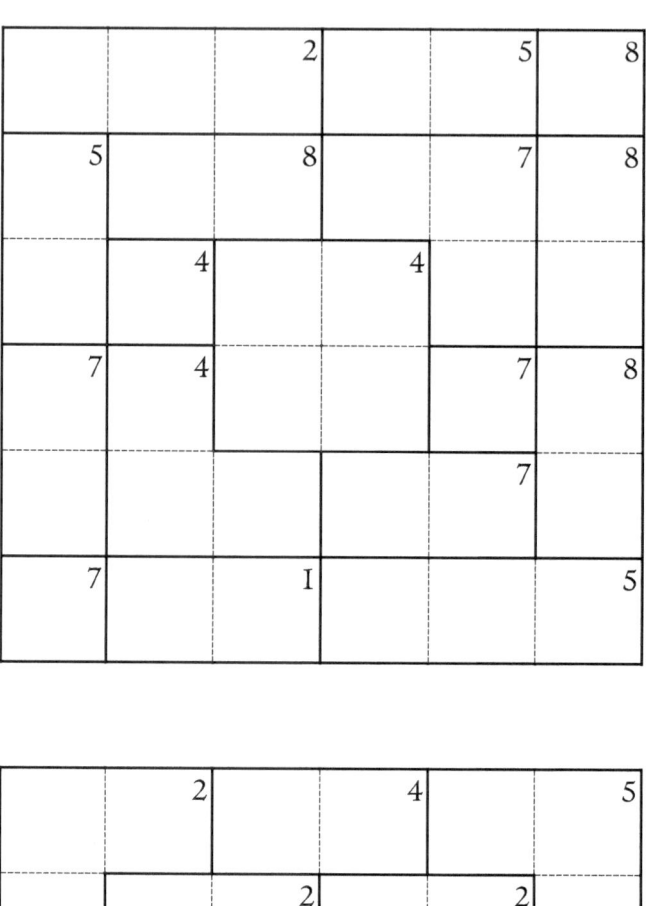

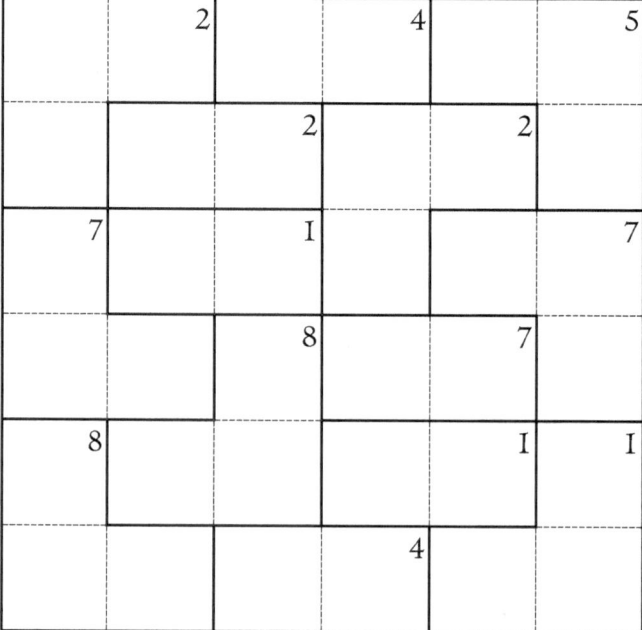

Fig. 3.46 Two puzzles over the group $U(9)$

3.3 Other Abelian Groups

Since we've seen that many groups are all basically equivalent, one might conjecture that every group of four elements is isomorphic to $\mathbb{Z}_4$. This would mean that for any order four group, there is some way of reordering those elements and mapping them over to $\{0, 1, 2, 3\}$ to arrive at the same group operation table as $\mathbb{Z}_4$. Is this always the case?

Consider the group $U(8)$. Since eight is two cubed, we have to remove all the even numbers to form our group of units. We are left with the set $\{1, 3, 5, 7\}$ and the table shown in Fig. 3.47.

We can see that 1 serves as an identity, the product of any two non-identity elements is the third non-identity element, and perhaps most noticeably, every element squared is equal to the identity. We insist that we always use the same order horizontally as we do vertically when constructing group tables, regardless of which order we choose. This means that we can determine the values of the squares in any finite group by examining the diagonal from the top left to the bottom right. Due to its importance, we will refer to these entries as the *main diagonal*.

Think about the values along this diagonal. No matter how we reorder or rename our four elements, the values here will contain only the identity. Therefore, no reordering or renaming can make this table look anything like the one for $\mathbb{Z}_4$, where the values along the main diagonal are not all the same. We have stumbled on a truly different group of order four.

For another example, we could consider the group $U(12)$. Here, we have to throw out all multiples of 2 or 3, leaving us with the set $\{1, 5, 7, 11\}$. Since each of these numbers squared is exactly one more than a multiple of twelve, we again find that each element times itself is equivalent to one, and hence is its own inverse. This group is shown in Fig. 3.48, and is not isomorphic to $\mathbb{Z}_4$ for the same reasons as $U(8)$.

Fig. 3.47 A table for the group $U(8)$

$\times$	1	3	5	7
1	1	3	5	7
3	3	1	7	5
5	5	7	1	3
7	7	5	3	1

Fig. 3.48 A table for the group $U(12)$

$\times$	1	5	7	11
1	1	5	7	11
5	5	1	11	7
7	7	11	1	5
11	11	7	5	1

Fig. 3.49 A table for the group $U(12)$

$\times$	$(0,0)$	$(1,0)$	$(0,1)$	$(1,1)$
$(0,0)$	$(0,0)$	$(1,0)$	$(0,1)$	$(1,1)$
$(1,0)$	$(1,0)$	$(0,0)$	$(1,1)$	$(0,1)$
$(0,1)$	$(0,1)$	$(1,1)$	$(0,0)$	$(1,0)$
$(1,1)$	$(1,1)$	$(0,1)$	$(1,0)$	$(0,0)$

One idea we haven't yet explored is the possibility of making bigger groups by combining smaller ones. For our next group of order four, let's start with the collection of all pairs of elements in $\mathbb{Z}_2$. We will call this $\mathbb{Z}_2 \times \mathbb{Z}_2$. This set is composed of the four elements $(0, 0)$, $(0, 1)$, $(1, 0)$, and $(1, 1)$.

Now we only need to define some operation. How might we combine two elements of this set, each of which has two parts? Perhaps the easiest choice is to just add the elements individually in each slot. This means for any two elements (a, b) and (y, z), we would get

$$(a, b) + (y, z) = (a + y, b + z)$$

where both the $a + y$ and $b + z$ computations take place in $\mathbb{Z}_2$. This means each slot of the output will also be in $\mathbb{Z}_2$, which shows that if we add two elements from $\mathbb{Z}_2 \times \mathbb{Z}_2$, we get another element from the same set. The identity will be $(0, 0)$, the inverse of each element will be itself, and because $\mathbb{Z}_2$ is associative, $\mathbb{Z}_2 \times \mathbb{Z}_2$ will be as well. We've found another group of order four, forged together from two identical groups of order two (Fig. 3.49).

The starting groups don't need to be identical to make this trick work. We can take this type of construction even further by defining what we call the *direct product* of any two groups. If the product in some group G is called $\times_G$, and the product in H is called $\times_H$, we can form a new group on the set of all pairs (g, h), where g is in G and h is in H, by defining

$$(g_1, h_1) \times (g_2, h_2) = (g_1 \times_G g_2, h_1 \times_H h_2).$$

We can get our identity from the pair of identities from each of the groups, our inverses from pairing together the inverses of any pair of elements, and associativity will follow from associativity of the individual groups. This means the direct product of any two groups will always be group itself.

We can perform a similar construction on triplets from three groups, and even extend further than that. This allows us to construct all sorts of new groups, though right now, we still have more groups of order four to introduce.

For our next example, we need to introduce a new definition. Given any set, a *subset* of that set is another set for which all the elements are contained in the original set. For example, with the set $\{1, 2\}$ there are four possible subsets: $\{\}, \{1\}, \{2\},$ and $\{1, 2\}$. The first set listed here is the set of no elements and is

Fig. 3.50 A table for the group $\Delta(\{1,2\})$

$\triangle$	$\{\}$	$\{1\}$	$\{2\}$	$\{1,2\}$
$\{\}$	$\{\}$	$\{1\}$	$\{2\}$	$\{1,2\}$
$\{1\}$	$\{1\}$	$\{\}$	$\{1,2\}$	$\{2\}$
$\{2\}$	$\{2\}$	$\{1,2\}$	$\{\}$	$\{1\}$
$\{1,2\}$	$\{1,2\}$	$\{2\}$	$\{1\}$	$\{\}$

Fig. 3.51 A table for the Klein four-group V

$\times$	e	a	b	c
e	e	a	b	c
a	a	e	c	b
b	b	c	e	a
c	c	b	a	e

Fig. 3.52 Isomorphisms of $U(8)$, $U(12)$, $\mathbb{Z}_2 \times \mathbb{Z}_2$, and $\Delta(\{1,2\})$ onto V

$$
\begin{array}{llll}
1 \to e & 1 \to e & (0,0) \to e & \{\} \to e \\
3 \to a & 5 \to a & (0,1) \to a & \{1\} \to a \\
5 \to b & 7 \to b & (1,0) \to b & \{2\} \to b \\
7 \to c & 11 \to c & (1,1) \to c & \{1,2\} \to c
\end{array}
$$

trivially a subset. Every element in it is also in $\{1,2\}$, simply because there are no elements in it. This set of no objects stands out, so we give it a special name, calling it the *empty set*.

We need some sort of operation on these four subsets to make it into a group, so let's define the *symmetric difference* of two sets to be the elements that are in exactly one of the two sets. Equivalently, we could say this is the set of elements in one or the other, but not both. Denoting this operation by the symbol "$\triangle$" we can make the table shown in Fig. 3.50.

We label this symmetric difference group $\Delta(\{1,2\})$, a name that gives us information about both the operation and the underlying set. All of the groups introduced in this section, including this one, are equivalent to the group shown in Fig. 3.51, which we call the *Klein four-group* and choose to denote by the letter V.[9] Maps giving these isomorphisms are shown in Fig. 3.52. These are not the only maps that work. There is so much symmetry in these groups that any relabeling of elements that sends the group identity to the identity of V will be an isomorphism.

None of these examples are cyclic. If we take any non-identity element g and repeatedly apply the operation to it with itself, we only get g and the identity, since g is its own inverse. If we do take the identity element and repeatedly apply the operation to it with itself, we get even less.

---

[9] The letter V comes from the word *Vierergruppe*. This is a German word that means "four-group" and the name given to this group by the mathematician Felix Klein [11].

Fig. 3.53 The eight subsets
of {1, 2, 3} organized by size
and inclusion

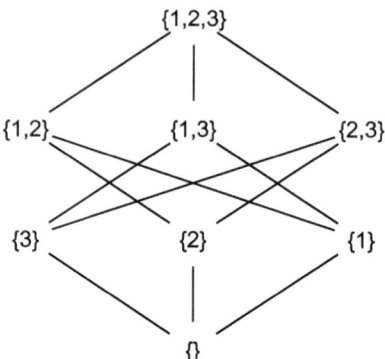

The structure of these groups leads to some interesting implications here. In puzzles over these groups, a clue cannot be equal to the identity in a cage of two cells. If it were, both elements inside the cage would need to be equal, violating the Latin square condition. Similarly, no one-by-three cage can contain its clue in one of its cells. If it did, the other two elements would need their product to be the identity, forcing those elements to be equal and again violating the Latin square condition. This restriction does not hold for all three-cell cages as L-shaped cages may have two elements which are equal.

For any set of two elements $\{a, b\}$ we could form the group $\triangle(\{a, b\})$, and it would be isomorphic to $\triangle(\{1, 2\})$. What would be happen if we used the same operation $\triangle$ on a three element set? Starting with $\{1, 2, 3\}$, we can construct a new group, $\triangle(\{1, 2, 3\})$, on the set of all eight possible subsets shown in Fig. 3.53. The empty set will still be the identity and every element is still its own inverse, because the symmetric difference of any set with itself results in the empty set.

This is our second group of order eight and it is distinct from groups we've seen so far. To illustrate this, consider that in $\mathbb{Z}_8$, not every element is its own inverse. In fact, the only elements in $\mathbb{Z}_8$ that are their own inverses are 0 and 4. Consequently, the identity appears only twice along the main diagonal in any table of a group isomorphic to $\mathbb{Z}_8$, whereas the identity must appear in every spot down this diagonal in any group isomorphic to $\triangle(\{1, 2, 3\})$. This makes it clear the two groups are not equivalent.

The group $\triangle(\{1, 2, 3\})$ is isomorphic to two other groups that we can already define using our existing terminology. The group $U(24)$, consisting of the elements

$$\{1, 5, 7, 11, 13, 17, 19, 23\}$$

is one such equivalent group. The direct product $\mathbb{Z}_2 \times \mathbb{Z}_2 \times \mathbb{Z}_2$, consisting of all triples where each element is in $\mathbb{Z}_2$, is isomorphic to this group as well.

In any group where every element is its own inverse, something very special happens: the solution to $ax = b$ is simply $x = ab$. To see this, simply multiply both sides on the left by a^{-1}. The left hand side is then

$$a^{-1}(ax) = (a^{-1}a)x = ex = x$$

and the right hand side is

$$a^{-1}b = ab.$$

In some sense this makes "division" the same as multiplication.

For an example of this in use, to solve

$$\{1, 2\}x = \{2, 3\}$$

one way is to simply take

$$x = \{1, 2\}\Delta\{2, 3\} = \{1, 3\}.$$

Another way is to ask what we need to add and remove in order to make one set into another. In this case to get to $\{2, 3\}$ from $\{1, 2\}$ we need to remove a 1 and add a 3 so the answer is $\{1, 3\}$. Both of these tricks make this group particularly nice to work in once one gets used to them, so we include several examples of puzzles over $\Delta(\{1, 2, 3\})$.

We have a cyclic group of order eight, and now these three new equivalent groups of order eight, but there is yet another abelian group with eight elements out there that is not isomorphic to either of these. The order eight groups $U(15)$, $U(16)$, $U(20)$ and $U(30)$ are all equivalent to each other, but are of a new type that we have not considered yet.

We choose to focus on $U(16)$ because its set of numbers is particularly easy to remember, consisting of all odd positive numbers less than 16. Its table is given in Fig. 3.54. To see that this group is not isomorphic to any of the previous groups, we present a structural difference, again involving the main diagonal. Unlike $\Delta(\{1, 2, 3\})$ where everything is its own inverse, and $\mathbb{Z}_8$ where exactly two elements are their own inverses, in $U(16)$, exactly four of the elements are their own inverse: $1, 7, 9$, and 15. No relabeling will change the number of times the identity appears on the main diagonal, so this group cannot be equivalent to the other two. This group is also not cyclic. In $U(16)$, every element raised to the fourth power is the identity, so no repeated application of the operation to one element can give us the entire eight-element group.

One might notice that we would also get abelian groups of order 8 by taking $\mathbb{Z}_2 \times \mathbb{Z}_4$ or $\mathbb{Z}_4 \times \mathbb{Z}_2$. This is true, and both of those constructions lead to groups isomorphic to $U(16)$. Since it is easier to write single numbers in cells than pairs, we stick to $U(16)$ when representing this collection of equivalent groups.

So far we've constructed all of our groups modulo n focusing only on non-negative integers. This type of arithmetic extends perfectly into the negatives as

Fig. 3.54 A group operation table for the group $U(16)$

$\times$	1	3	5	7	9	11	13	15
1	1	3	5	7	9	11	13	15
3	3	9	15	5	11	1	7	13
5	5	15	9	3	13	7	1	11
7	7	5	3	1	15	13	11	9
9	9	11	13	15	1	3	5	7
11	11	1	7	13	3	9	15	5
13	13	7	1	11	5	15	9	3
15	15	13	11	9	7	5	3	1

well, by letting -1 be equivalent to $n - 1$, -2 be equivalent to $n - 2$, and so on. In arithmetic modulo sixteen, the elements $9, 11, 13$ and 15 are equivalent to $-7, -5, -3,$ and -1 respectively. It may be easier to work with the set

$$\{1, -1, 3, -3, 5, -5, 7, -7\}$$

instead of only using positive odd numbers. To show how negatives can help, notice that once we know that

$$7 \times 5 = 3$$

we immediately also know that

$$7 \times -5 = -3,$$

$$-7 \times 5 = -3,$$

and

$$-7 \times -5 = 3.$$

This form of manipulation of negatives can be used to turn every single product into four rules. The elements ± 1 and ± 7 are their own inverses, and since 3 and -5 are inverses of each other, we know -3 and 5 must be inverses of each other as well. We encourage the reader to work over this set instead and thus we replace the $9, 11, 13$ and 15 in the $U(16)$ puzzles with their negative counterparts.

Puzzle Rules

Fill in the cells so that each element of the abelian group appears exactly once in every row and column. The clues in each cage indicate the product of the values from their cells. The following tables (Figs. 3.55, 3.56, 3.57, 3.58, and 3.59) may be helpful for solving these puzzles (Figs. 3.60, 3.61, 3.62, 3.63, 3.64, 3.65, 3.66, 3.67, 3.68, 3.69, 3.70, 3.71, 3.72, 3.73, 3.74, 3.75, 3.76, 3.77, and 3.78).

Δ	{}	{1}	{2}	{3}	{2, 3}	{1, 3}	{1, 2}	{1, 2, 3}
{}	{}	{1}	{2}	{3}	{2, 3}	{1, 3}	{1, 2}	{1, 2, 3}
{1}	{1}	{}	{1, 2}	{1, 3}	{1, 2, 3}	{3}	{2}	{2, 3}
{2}	{2}	{1, 2}	{}	{2, 3}	{3}	{1, 2, 3}	{1}	{1, 3}
{3}	{3}	{1, 3}	{2, 3}	{}	{2}	{1}	{1, 2, 3}	{1, 2}
{2, 3}	{2, 3}	{1, 2, 3}	{3}	{2}	{}	{1, 2}	{1, 3}	{1}
{1, 3}	{1, 3}	{3}	{1, 2, 3}	{1}	{1, 2}	{}	{2, 3}	{2}
{1, 2}	{1, 2}	{2}	{1}	{1, 2, 3}	{1, 3}	{2, 3}	{}	{3}
{1, 2, 3}	{1, 2, 3}	{2, 3}	{1, 3}	{1, 2}	{1}	{2}	{3}	{}

Fig. 3.55 The group operation table for the group $\Delta(\{1, 2, 3\})$

Fig. 3.56 All possibilities for two cell cages in $\Delta(\{1, 2, 3\})$

{2, 3}	{1, 3}	{1, 2}	{1, 2, 3}
{}Δ {2, 3}	{}Δ {1, 3}	{}Δ {1, 2}	{}Δ {1, 2, 3}
{2}Δ {3}	{1}Δ {3}	{1}Δ {2}	{1}Δ {2, 3}
{1, 2}Δ {1, 3}	{1, 2}Δ {2, 3}	{1, 3}Δ {2, 3}	{2}Δ {1, 3}
{1}Δ {1, 2, 3}	{2}Δ {1, 2, 3}	{3}Δ {1, 2, 3}	{3}Δ {1, 2}

Fig. 3.57 Inverses of elements in $\Delta(\{1, 2, 3\})$

a	{}	{1}	{2}	{3}	{2, 3}	{1, 3}	{1, 2}	{1, 2, 3}
a^{-1}	{}	{1}	{2}	{3}	{2, 3}	{1, 3}	{1, 2}	{1, 2, 3}

Fig. 3.58 The group operation table for the group $U(16)$ over the set $\{1, -1, 3, -3, 5, -5, 7, -7\}$

$\times$	1	3	5	7	-7	-5	-3	-1
1	1	3	5	7	-7	-5	-3	-1
3	3	-7	-1	5	-5	1	7	-3
5	5	-1	-7	3	-3	7	1	-5
7	7	5	3	1	-1	-3	-5	-7
-7	-7	-5	-3	-1	1	3	5	7
-5	-5	1	7	-3	3	-7	-1	5
-3	-3	7	1	-5	5	-1	-7	3
-1	-1	-3	-5	-7	7	5	3	1

Fig. 3.59 All possibilities for two cell cages in $U(16)$

1	3	5	7	-7	-5	-3	-1
$3 \cdot -5$	$1 \cdot 3$	$1 \cdot 5$	$1 \cdot 7$	$1 \cdot -7$	$1 \cdot -5$	$1 \cdot -3$	$1 \cdot -1$
$5 \cdot -3$	$-1 \cdot -3$	$-1 \cdot -5$	$-1 \cdot -7$	$-1 \cdot 7$	$-1 \cdot 5$	$-1 \cdot 3$	$3 \cdot 5$
	$5 \cdot 7$	$3 \cdot 7$	$3 \cdot -3$		$-3 \cdot 7$	$5 \cdot -7$	$-3 \cdot -5$
	$-5 \cdot -7$	$-3 \cdot -7$	$5 \cdot -5$		$3 \cdot -7$	$-5 \cdot 7$	$7 \cdot -7$

Fig. 3.60 Inverses of elements in $U(16)$

a	1	3	5	7	-7	-5	-3	-1
a^{-1}	1	-5	-3	7	-7	3	5	-1

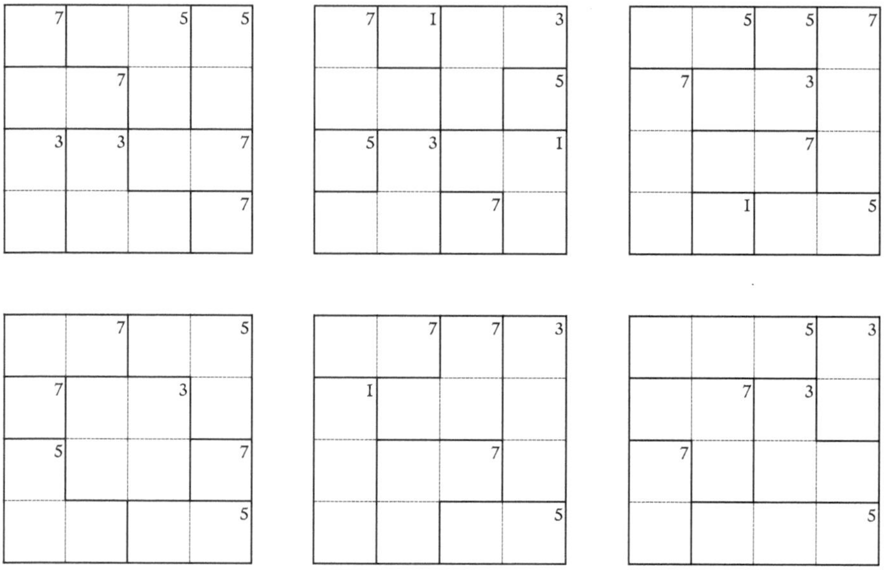

Fig. 3.61 Six puzzles over the group $U(8)$

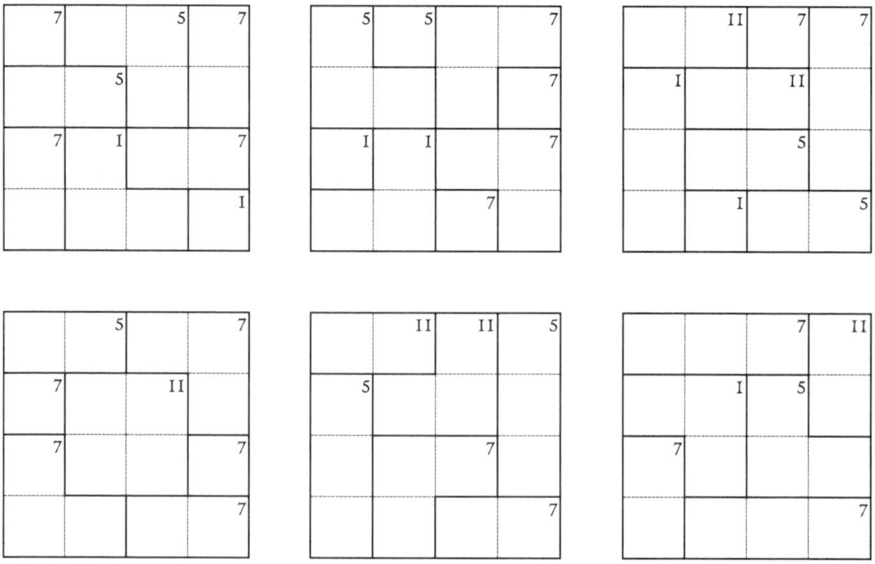

Fig. 3.62 Six puzzles over the group $U(12)$

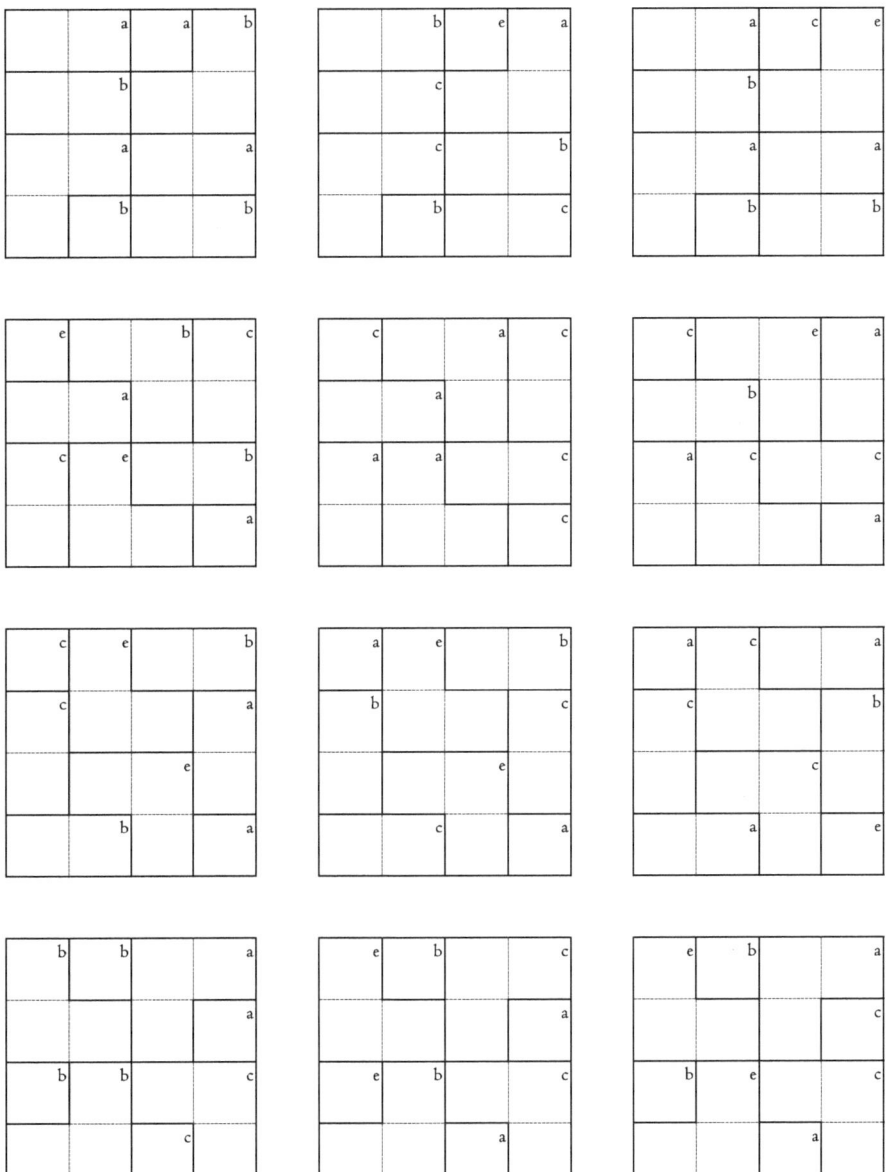

Fig. 3.63 Twelve puzzles over the Klein four-group V

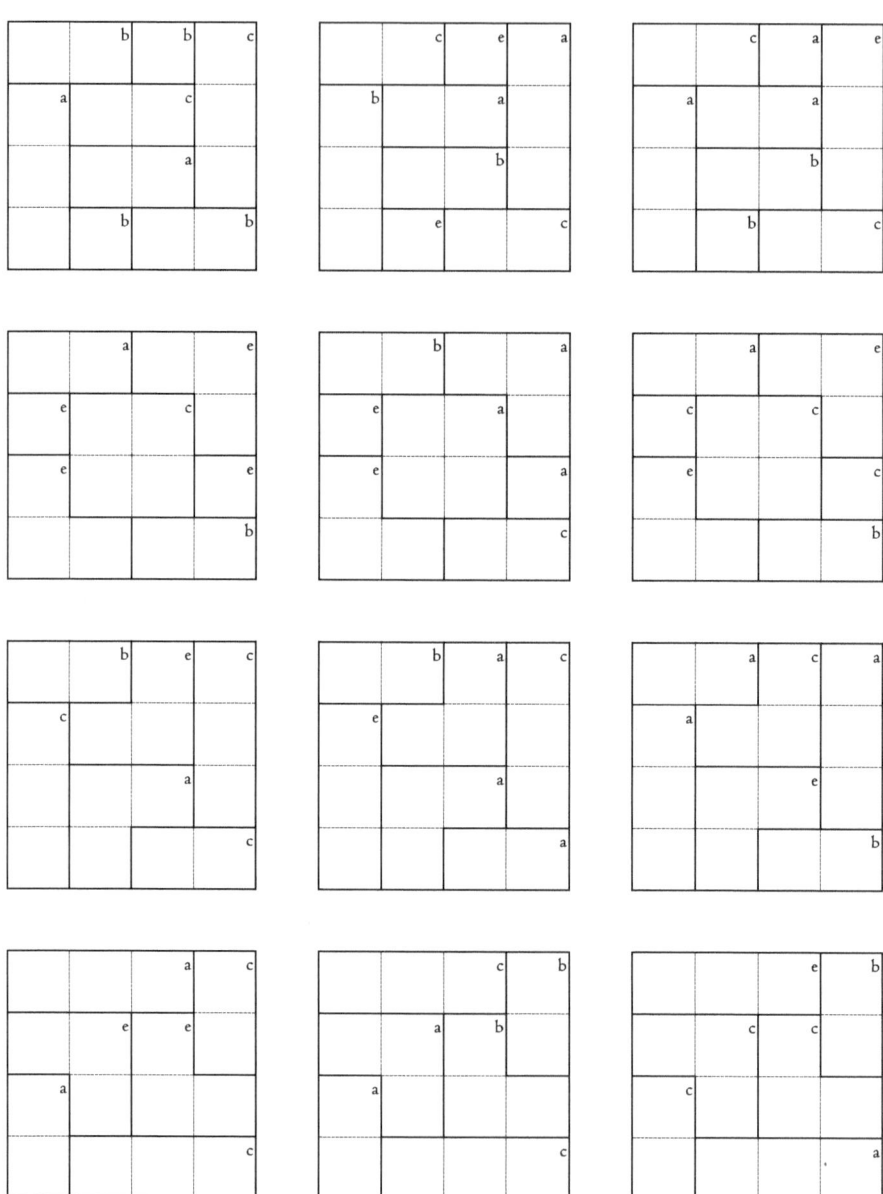

Fig. 3.64 Twelve puzzles over the Klein four-group *V*

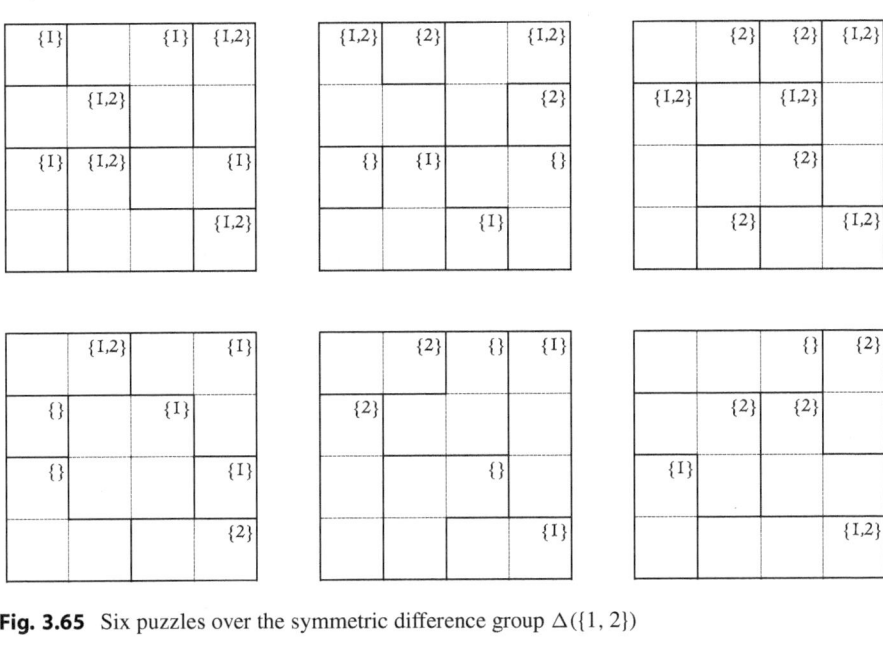

Fig. 3.65 Six puzzles over the symmetric difference group $\Delta(\{1, 2\})$

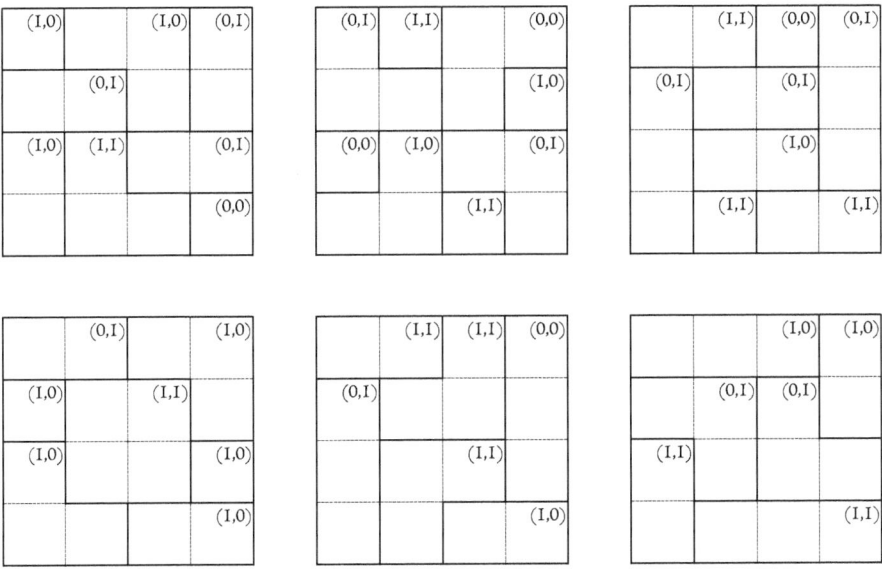

Fig. 3.66 Six puzzles over the group $\mathbb{Z}_2 \times \mathbb{Z}_2$

{1,2}	{1,2}			{3}	{1,3}		{1,2}
		{1,2}	{2}	{3}	{1}	{1,2,3}	{1,2}
{1,2,3}		{1,2,3}		{1,2}			{1}
{}	{}		{3}			{2,3}	
		{3}	{3}			{3}	{1,2,3}
	{1,3}	{1}		{2}		{1,3}	{1}
{3}			{2}			{1,2}	{3}
	{2,3}	{3}			{1,3}	{1,3}	

Fig. 3.67 A puzzle over the symmetric difference group $\Delta(\{1, 2, 3\})$

{1,2,3}		{1,2}		{1}		{1}	{3}
{2}	{2}		{3}		{1}	{1,2}	{1,2,3}
	{1,3}	{2,3}	{3}	{2,3}	{}	{1,3}	
{1,3}							{2,3}
	{1,3}		{1}		{3}	{3}	
{3}		{1}			{2}		{2,3}
	{1,2}		{2}		{2,3}	{3}	
{3}		{1,2,3}		{1,2}		{1,2}	{1,2}

Fig. 3.68 A puzzle over the symmetric difference group $\Delta(\{1, 2, 3\})$

		{1,2}		{1}		{1,2}	{2}
{2,3}	{1,3}		{2}			{2,3}	
	{2,3}		{2,3}	{1,2,3}	{2}	{3}	
{2}			{2,3}				{3}
	{1}		{1,2}	{1,2,3}	{1,3}	{3}	
{}		{2,3}					{1,3}
			{2}		{1,3}	{}	
		{1,2}		{2,3}			{}

Fig. 3.69 A puzzle over the symmetric difference group $\Delta(\{1, 2, 3\})$

{2}	{3}	{1,2,3}	{1}	{2,3}	{1,3}		{2,3}
							{1,3}
	{1,3}	{1,2}		{1,3}	{3}		{1,2,3}
	{2,3}	{1,2}	{2,3}	{}	{1,2}		{2}
	{1,3}		{1,2}	{1}			{1,3}
	{1,2,3}	{}		{1,3}	{1,2}		{1}
	{2,3}	{1,2,3}	{2,3}	{1,3}	{1}	{1}	{1,2}
	{1}						

Fig. 3.70 A puzzle over the symmetric difference group $\Delta(\{1, 2, 3\})$

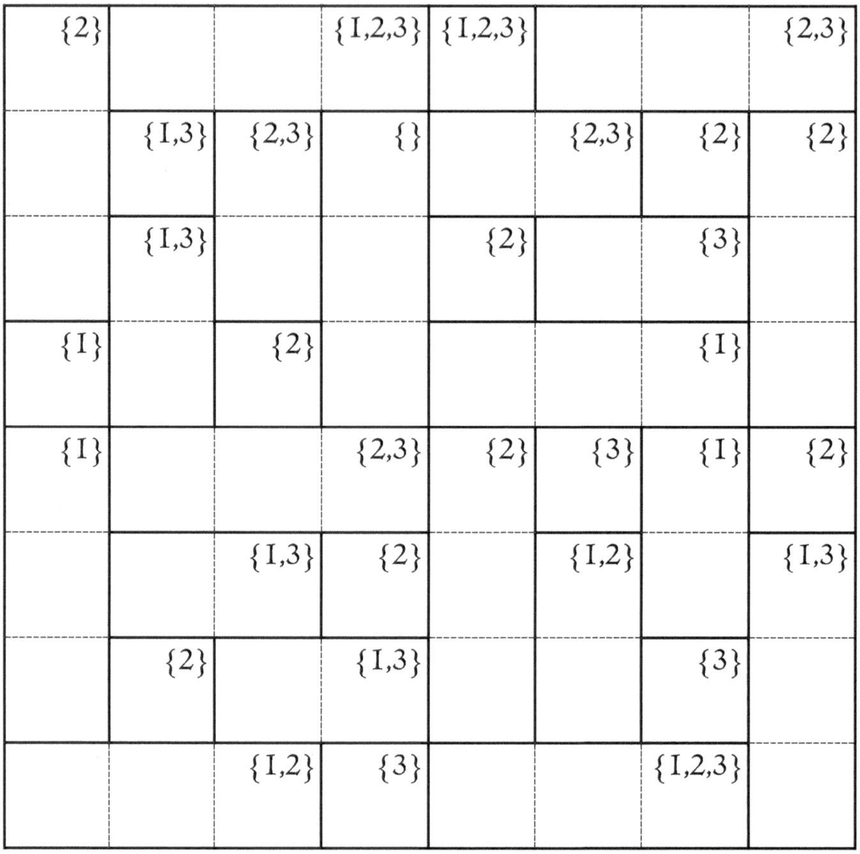

{2}			{1,2,3}	{1,2,3}			{2,3}
	{1,3}	{2,3}	{}		{2,3}	{2}	{2}
	{1,3}			{2}		{3}	
{1}		{2}				{1}	
{1}			{2,3}	{2}	{3}	{1}	{2}
		{1,3}	{2}		{1,2}		{1,3}
	{2}		{1,3}			{3}	
		{1,2}	{3}			{1,2,3}	

Fig. 3.71 A puzzle over the symmetric difference group $\Delta(\{1, 2, 3\})$

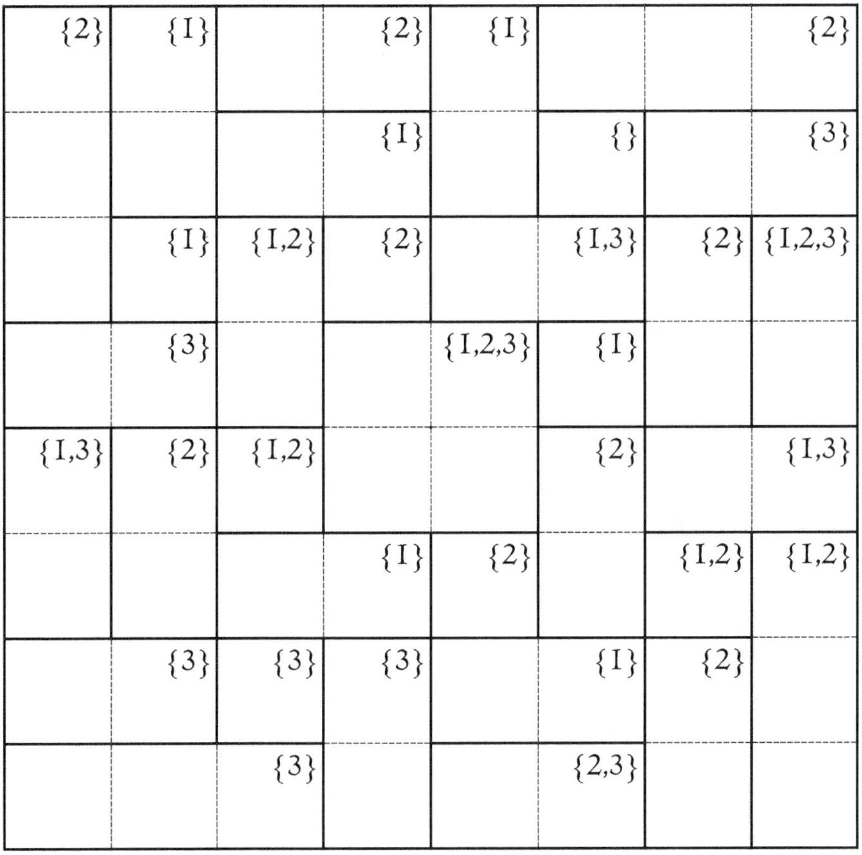

{2}	{1}		{2}	{1}			{2}
			{1}		{}		{3}
	{1}	{1,2}	{2}		{1,3}	{2}	{1,2,3}
	{3}			{1,2,3}	{1}		
{1,3}	{2}	{1,2}			{2}		{1,3}
			{1}	{2}		{1,2}	{1,2}
	{3}	{3}	{3}		{1}	{2}	
		{3}			{2,3}		

Fig. 3.72 A puzzle over the symmetric difference group $\Delta(\{1, 2, 3\})$

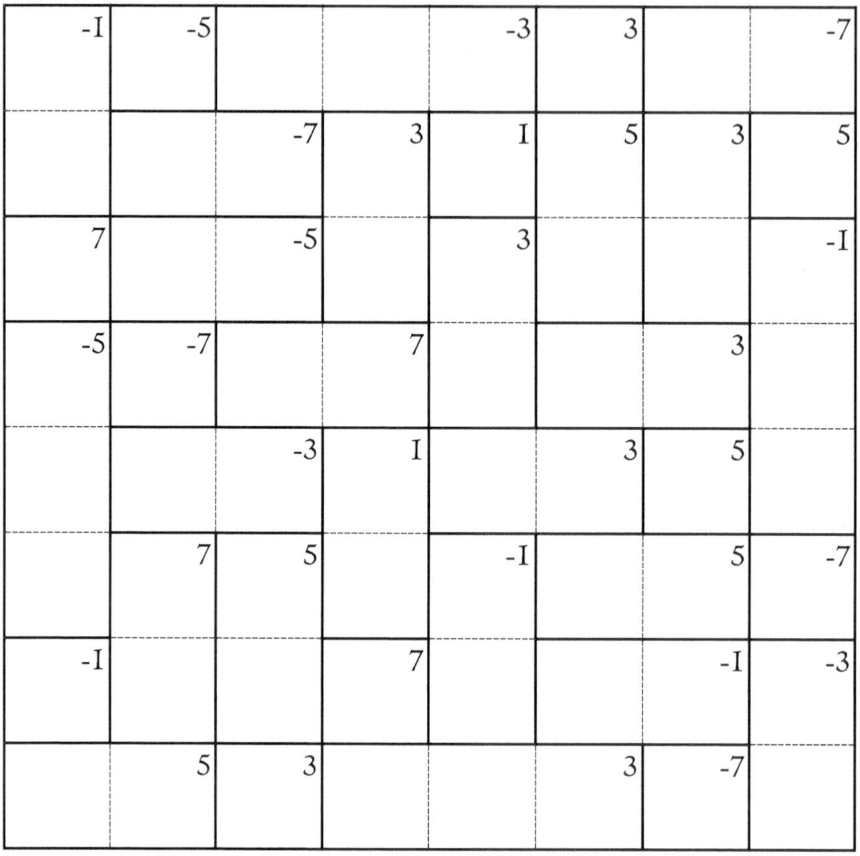

-1	-5			-3	3		-7
		-7	3	1	5	3	5
7		-5		3			-1
-5	-7		7			3	
		-3	1		3	5	
	7	5		-1		5	-7
-1			7			-1	-3
	5	3			3	-7	

Fig. 3.73 A puzzle over the group $U(16)$

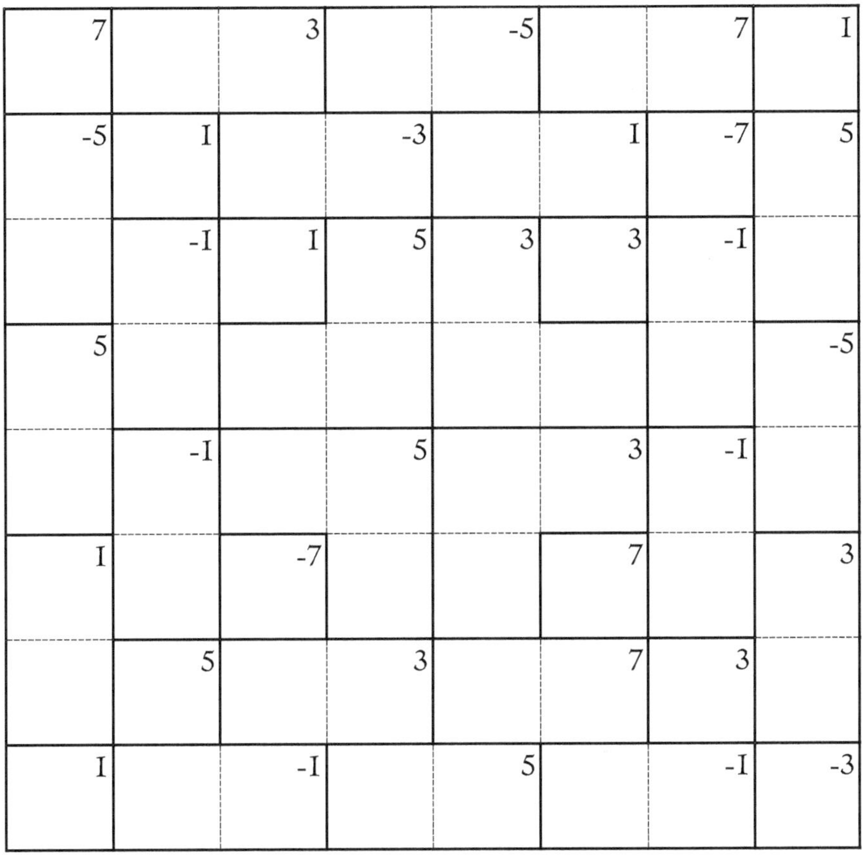

7		3		-5		7	I
-5	I		-3		I	-7	5
	-I	I	5	3	3	-I	
5							-5
	-I		5		3	-I	
I		-7			7		3
	5		3		7	3	
I		-I		5		-I	-3

Fig. 3.74 A puzzle over the group $U(16)$

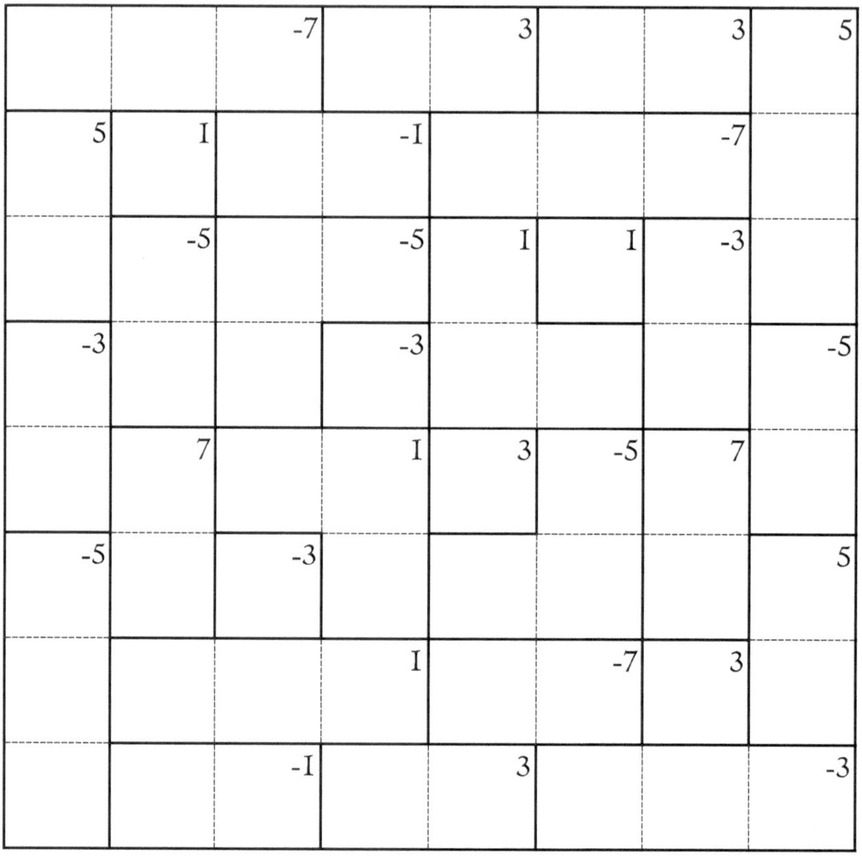

		-7		3		3	5
5	1		-1			-7	
	-5		-5	1	1	-3	
-3			-3				-5
	7		1	3	-5	7	
-5		-3					5
			1		-7	3	
		-1		3			-3

Fig. 3.75 A puzzle over the group $U(16)$

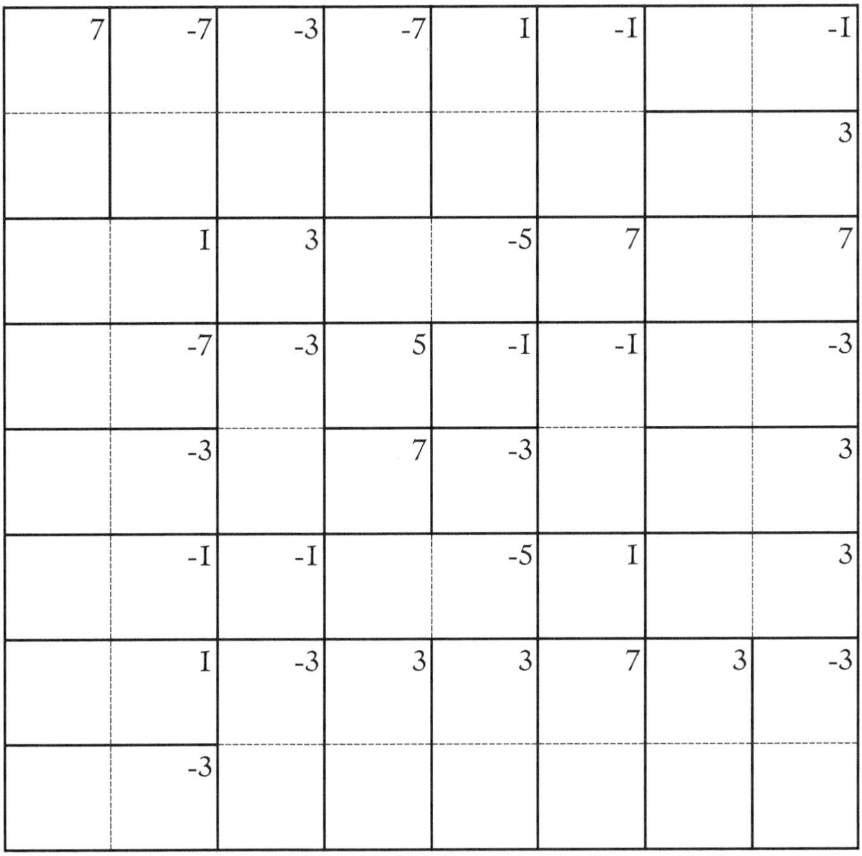

7	-7	-3	-7	I	-I		-I
							3
	I	3		-5	7		7
	-7	-3	5	-I	-I		-3
	-3		7	-3			3
	-I	-I		-5	I		3
	I	-3	3	3	7	3	-3
	-3						

Fig. 3.76 A puzzle over the group $U(16)$

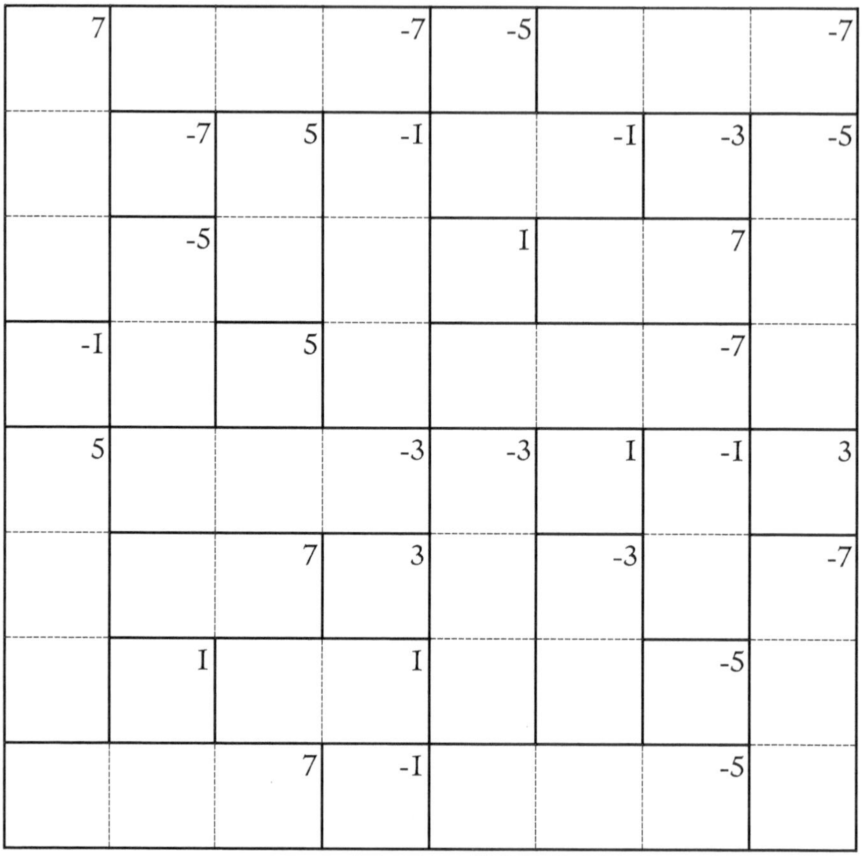

7				-7	-5			-7
	-7	5	-I			-I	-3	-5
	-5				I		7	
-I		5					-7	
5				-3	-3	I	-I	3
			7	3		-3		-7
		I		I			-5	
			7	-I			-5	

Fig. 3.77 A puzzle over the group $U(16)$

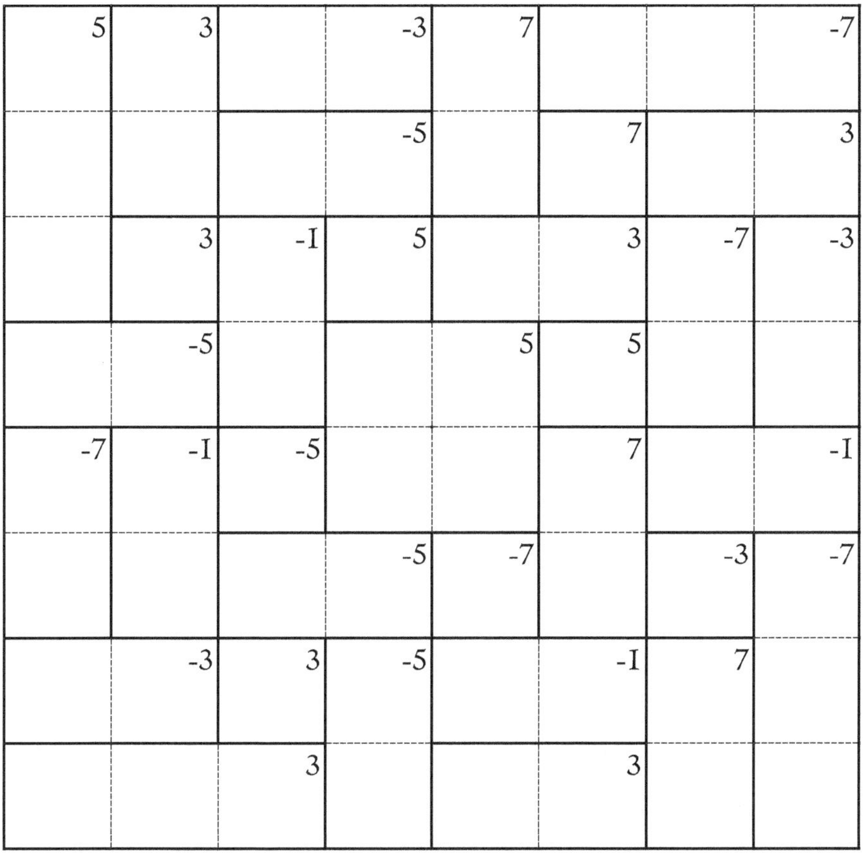

5	3		-3	7			-7
			-5		7		3
	3	-1	5		3	-7	-3
	-5			5	5		
-7	-1	-5			7		-1
			-5	-7		-3	-7
	-3	3	-5		-1	7	
		3			3		

Fig. 3.78 A puzzle over the group $U(16)$

Non-abelian Groups

4

4.1 Dihedral Groups

Suppose that we have a closed copy of a book in front of us. We place it on a desk facing us and proceed to rotate the book by sliding it according to a series of commands. We let r symbolize the process of rotating this book about its center $90°$ clockwise and define the product of two rotations to be the process of doing both one after another. Thus $r \cdot r$, which we can denote by r^2, amounts to rotating by $90°$ and then doing it again, for a final rotation of $180°$. Define r^n to be the act of applying r a total of n times, for all higher values of n. Let e be the process of not moving the book in any way. We can now perform any of the operations in the list

$$\{e, r, r^2, r^3, r^4, r^5, \cdots\}.$$

We want to somehow make this list into a group, with two sets of instructions being equal if they move the book from the same starting position to the same ending position. Both $r \cdot e$ and $e \cdot r$ leave the book in the same position as r, hence $r \cdot e = e \cdot r = r$. Notice also that r^4, which rotates $90°$ four times, leaves the book in the same position as e, and that r^5 leaves the book in the same position as r. Any collection involving four or more r's simplifies to one with three or fewer.[1] We can make a multiplication table for the only four distinct motions we can achieve, as shown in Fig. 4.1.

---

[1] If miniature versions of us were sitting on the book, they would certainly notice spinning more in r^5 than in r^1, but that makes no difference for where they end up, and thus no difference in this definition.

Fig. 4.1 A group of rotations

$\times$	e	r	r^2	r^3
e	e	r	r^2	r^3
r	r	r^2	r^3	e
r^2	r^2	r^3	e	r
r^3	r^3	e	r	r^2

Collections of motions like these are always associative, because applying three motions in sequence always yields the same result. Whether we apply the first motion and then the combination of the next two, or apply the combination of the first two together followed by the third, the outcome remains unchanged. All the elements are invertible, as $e^{-1} = e$, $r^{-1} = r^3$, $(r^2)^{-1} = r^2$ and $(r^3)^{-1} = r$. The element e acts as an identity, so all three rules are met, and what we have here is the structure of a group. Notice that the table here follows the exact same pattern as $\mathbb{Z}_4$ and the many other groups isomorphic to it that we have already examined.

We are now about to add something that will make this group quite different: a flip.[2] Let f be the motion that flips the book over an invisible line going through a point in the center of the book. It doesn't matter how one visualizes this line, so long as they do not allow the line to rotate with the book, but instead keep it in place as the book moves.

Consider the collection of instructions for motions, written as sequences of r's and f's, where r is as before and f is our new flip. As before, e is the instruction to not touch the book, and two instructions are considered equivalent if their end result leaves the book in the same position. Both rrr and ff are equal to the identity, because these transformations return the book to its original spot. Both of these instructions read the same from right to left, but we have to make a decision about what we want an instruction such as $rfrrf$ to mean.

We choose to read all instructions from right to left simply because we would like our notation to match the common one for compositions in pre-calculus and calculus. There,

$$f \circ g(x) = f(g(x))$$

and thus g is applied to x first, even though it is on the right. This means that here, the transformation rf will flip our book and then rotate it 90° clockwise. The transformation fr will rotate our book first, and then flip it. We encourage the reader to try this now and see that these two do not leave the book in the same position, and thus these transformations are not the same. The difference between these transformations is also illustrated in Fig. 4.2.

---

[2] Any miniature versions of us sitting on the book might want to get off now.

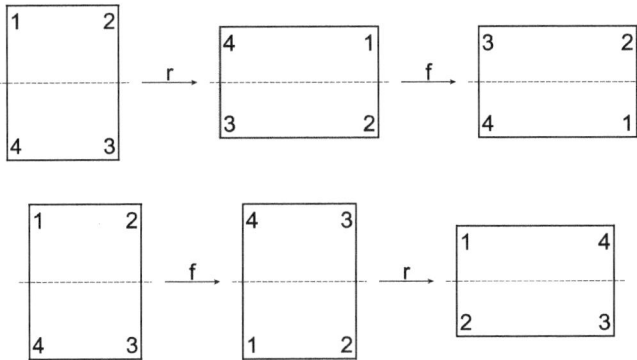

Fig. 4.2 The transformation rf affects our object differently than the transformation fr

The instructions rf and fr differ by a 180° rotation. We would need to rotate twice more to get between these two outcomes, thus what fr does is the same as the transformation $r^3 f$. This establishes the rule

$$fr = r^3 f.$$

We can use this rule to rewrite any instruction where an r appears to the right of an f, eventually moving all the r's to the left of all the f's. Therefore any expression can be transformed into one where all our rotations take place after any flips. For example, $rfrrf$ is equivalent to $rr^3 frf$, which equals frf. We can simplify further, as frf is the same as $r^3 ff$, which is equivalent to r^3.

If we reduce our instructions in this way, we are left with 0, 1, 2 or 3 rotations, which are all to the left of 0 or 1 flips. This leaves us exactly eight final possibilities after our reduction. Every set of instructions is equivalent to one from the following set

$$\{e, r, r^2, r^3, f, rf, r^2 f, r^3 f\}.$$

We include a picture of these eight transformations in Fig. 4.3.

We still need a product if we wish to make a group out of this set. If a and b are two sets of instructions, we multiply them by placing them side by side to make a new sequence ab. For example $rf \times r^2 f$ equals the sequence $rfr^2 f$, which we've already seen is equivalent to r^3. Notice that with this definition

$$r \times f \neq f \times r$$

as we've already seen that rf is a different transformation than fr. If we can show every transformation has an inverse, we will have found our first example of a non-abelian group!

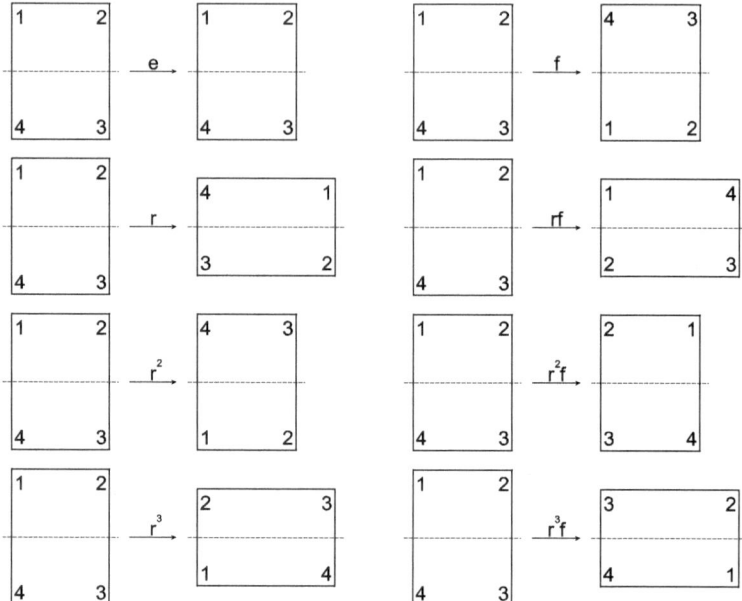

Fig. 4.3 A picture of eight motions

We only need to show that each of the eight transformations have another transformation we can apply that returns the book to its initial position. The elements e, r^2 and each of the four transformations containing an f each undo themselves when applied again. This leaves r and r^3, which each undo each other. Thus, we have just found a non-abelian group of order eight that we refer to from now on as D_4. The subscript here refers to the number of rotations required to get back to where we started, which in this case is the number 4.

There was no reason we had to choose a 90° rotation. We could have chosen 120°, to construct a structure where $r^3 = e$ or 72° to construct one where $r^5 = e$. We could continue to use a book or any other flippable object and then try to construct the groups D_3 and D_5 in a similar fashion. There is also a slightly different approach that eliminates the idea of flips and rotations altogether. This is the route we mean to take next.

First, we fix some positive integer n and then consider the set of all words of any size that we can make with the letters r and f. By this, we mean all collections of adjacent letters of finite length, and not just collections that we would find in a dictionary.[3] The collection of all words in these letters is the infinite set

$$\{r, f, rr, rf, fr, ff, rrr, rrf, rfr, rff, frr, frf, ffr, fff, \cdots\}.$$

---

[3] With only r and f one would need a strange dictionary to find these words.

Allow e to be a symbol that represents an empty word, or word with no letters. We then define multiplication on these words through juxtaposition, simply placing them next to each other to form a new one. Thus

$$rff \times rr = rffrr,$$

$$e \times rfr = rfr,$$

and

$$e \times e = e.$$

Next, consider the set of transformations of these words where we are allowed to replace r^n with e, f^2 with e and fr with $r^{n-1}f$ at any place in the word where these letters appear. We can write

$$r^n \to e,$$

$$f^2 \to e,$$

and

$$fr \to r^{n-1}f$$

for short. We define two words to be the same if we can turn them into the same result using these transformations.

Given any word, we can repeatedly use the third rule to move all our r's to the left, and then use the other two rules to ensure there is at most one f and at most $n - 1$ r's. Every word ends up equivalent to exactly one of the elements in the $2n$ element set

$$\{e, r, r^2, \cdots, r^{n-1}, f, rf, r^2f, \cdots, r^{n-1}f\}.$$

This construction gives us an associative structure with identity e, where the inverse of r^k is r^{n-k}, and the inverse of $r^k f$ is itself. We therefore have arrived at a group of order $2n$ which we can call the *dihedral group* of order n, or D_n.[4]

Whenever we specify a group using words on a set of letters with some system of reductions, we call this a *group presentation*. We could have defined all the groups we have seen so far, as well as all the groups still to come, in this fashion, by carefully choosing the right letters and reductions.

---

[4] The earliest writing on these groups seems to appear in a book of Felix Klein [11]. The name comes from the word *dihedron*, which refers to a two-sided or flattened polyhedron.

In this dihedral case, the reductions for our words come from real facts about motions of objects in space. This should reassure us that words must always end up reduced in a unique way. No one word could lead to two different reduced forms, as each reduced form represents a different physical motion, and our initial word represents only one such motion.

If the reductions were less obvious and not grounded in some physical reality, it might not be clear when two words were actually equivalent. Words will often have many different places where we can start applying our reductions. Depending on our choices during the reduction process, we could end up with very different-looking outcomes, and thus many different-looking words might represent the same group element. There is no easy way in general to tell exactly how many distinct elements there will be. In fact, there is no efficient method for even just telling when a presented group has all of its elements the same!

By moving from rotations to words, we have turned a very visual construction into something more abstract, but this new definition contains all the information that we need in order to tackle puzzles over these groups without us having to resort to moving any objects around. There is still one issue that remains for puzzle construction. These groups are not abelian. How do we deal with the issue of the order of elements, in terms of determining when the product of entries in the cells of a given cage meet the clue for that cage?

There are several ways one might deal with this issue. One possibility is when making clues for a puzzle, we can consider all the possible outcomes from placing elements in different orders, and pick the smallest outcome according to some ordering. However, this method has its own weaknesses. Not only do we need to pick an ordering, but the distribution of clues is skewed greatly by which ordering we pick, and the possible solutions for a cage with a given clue ends up feeling more dependent on our arbitrary ordering than anything else.

What we do instead here, and in the other non-abelian groups to come, is simply insist that all products are taken in a particular direction. For rectangular $1 \times n$ or $n \times 1$ cages, we always work from left to right and top to bottom. For L-shaped cages, we always work clockwise. This sort of method is perhaps the easiest choice for a puzzler to deal with, instead of having to remember some ordering.

Since clues in a puzzle are small, and exponents inside a clue would be smaller, we choose to write out rr, rrr, rrf and $rrrf$ in the following puzzles when they appear inside a clue, instead of using the conventional notation r^2, r^3, $r^2 f$ and $r^3 f$. This is done only for appearance, and the reader is encouraged to fill in the cells with whichever notation they like best.

When $a \times b = b \times a$ for two distinct elements in a group, we say that those two elements *commute* with each other. Even in non-abelian groups where many elements do not commute with each other, there are always some elements which do. In both D_3 and D_4, the identity commutes with all other elements, and powers of r commute with all other powers of r. In D_4, the element r^2 commutes with all other elements in the group, though this is not the case in D_3. Recalling these facts can help with the puzzles ahead.

Puzzle Rules

Fill in the cells so that each element of the group D_3 or D_4 appears exactly once in every row and column. The clues in each cage indicate the product of the cells when taken top to bottom or left to right for $n \times 1$ or $1 \times n$ shaped cages, and when taken clockwise for L-shaped cages. The following tables (Figs. 4.4, 4.5, 4.6, 4.7, 4.8, and 4.9) may be helpful for solving these puzzles (Figs. 4.10, 4.11 4.12, 4.13, 4.14, 4.15, 4.16, 4.17, 4.18, 4.19, 4.20, and 4.21).

Fig. 4.4 The group operation table for the group D_3

$\times$	e	r	rr	f	rf	rrf
e	e	r	rr	f	rf	rrf
r	r	rr	e	rf	rrf	f
rr	rr	e	r	rrf	f	rf
f	f	rrf	rf	e	rr	r
rf	rf	f	rrf	r	e	rr
rrf	rrf	rf	f	rr	r	e

Fig. 4.5 All possibilities for two cell cages in D_3

e	r	rr	f	rf	rrf
$r \cdot rr$	$e \cdot r$	$e \cdot rr$	$e \cdot f$	$e \cdot rf$	$e \cdot rrf$
$rr \cdot r$	$r \cdot e$	$rr \cdot e$	$r \cdot rrf$	$r \cdot f$	$r \cdot rf$
	$f \cdot rrf$	$f \cdot rf$	$rr \cdot rf$	$rr \cdot rrf$	$rr \cdot f$
	$rf \cdot f$	$rf \cdot rrf$	$f \cdot e$	$f \cdot rr$	$f \cdot r$
	$rrf \cdot rf$	$rrf \cdot f$	$rf \cdot r$	$rf \cdot e$	$rf \cdot rr$
			$rrf \cdot rr$	$rrf \cdot r$	$rrf \cdot e$

Fig. 4.6 Inverses of elements in D_3

a	e	r	rr	f	rf	rrf
a^{-1}	e	rr	r	f	rf	rrf

×	e	r	rr	rrr	f	rf	rrf	$rrrf$
e	e	r	rr	rrr	f	rf	rrf	$rrrf$
r	r	rr	rrr	e	rf	rrf	$rrrf$	f
rr	rr	rrr	e	r	rrf	$rrrf$	f	rf
rrr	rrr	e	r	rr	$rrrf$	f	rf	rrf
f	f	$rrrf$	rrf	rf	e	rrr	rr	r
rf	rf	f	$rrrf$	rrf	r	e	rrr	rr
rrf	rrf	rf	f	$rrrf$	rr	r	e	rrr
$rrrf$	$rrrf$	rrf	rf	f	rrr	rr	r	e

Fig. 4.7 The group operation table for the group D_4

e	r	rr	rrr	f	rf	rrf	$rrrf$
$r \cdot rrr$	$e \cdot r$	$e \cdot rr$	$e \cdot rrr$	$e \cdot f$	$e \cdot rf$	$e \cdot rrf$	$e \cdot rrrf$
$rrr \cdot r$	$r \cdot e$	$rr \cdot e$	$r \cdot rr$	$r \cdot rrrf$	$r \cdot f$	$r \cdot rf$	$r \cdot rrf$
	$rr \cdot rrr$	$f \cdot rrf$	$rr \cdot r$	$rr \cdot rrf$	$rr \cdot rrrf$	$rr \cdot f$	$rr \cdot rf$
	$rrr \cdot rr$	$rf \cdot rrrf$	$rrr \cdot e$	$rrr \cdot rf$	$rrr \cdot rrf$	$rrr \cdot rrrf$	$rrr \cdot f$
	$f \cdot rrrf$	$rrf \cdot f$	$f \cdot rf$	$f \cdot e$	$f \cdot rrr$	$f \cdot rr$	$f \cdot r$
	$rf \cdot f$	$rrrf \cdot rf$	$rf \cdot rrf$	$rf \cdot r$	$rf \cdot e$	$rf \cdot rrr$	$rf \cdot rr$
	$rrf \cdot rf$		$rrf \cdot rrrf$	$rrf \cdot rr$	$rrf \cdot r$	$rrf \cdot e$	$rrf \cdot rrr$
	$rrrf \cdot rrf$		$rrrf \cdot f$	$rrrf \cdot rrr$	$rrrf \cdot rr$	$rrrf \cdot r$	$rrrf \cdot e$

Fig. 4.8 All possibilities for two cell cages in D_4

Fig. 4.9 Inverses of elements in D_4

a	e	r	rr	rrr	f	rf	rrf	$rrrf$
a^{-1}	e	rrr	rr	r	f	rf	rrf	$rrrf$

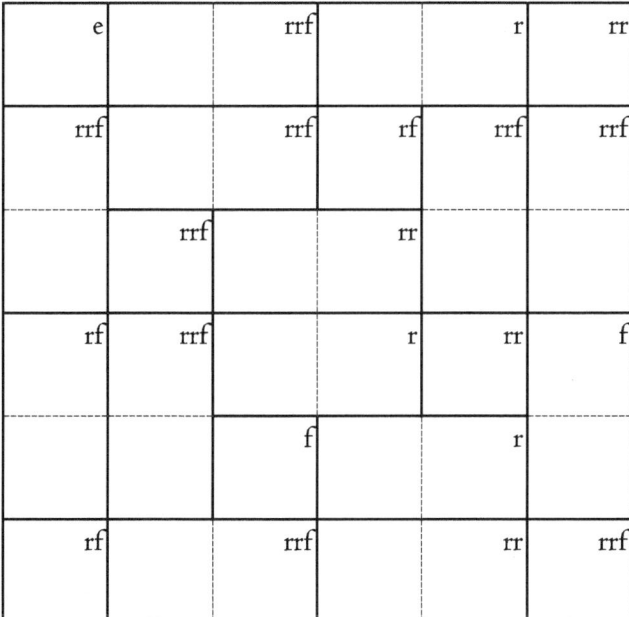

Fig. 4.10 Two puzzles over the dihedral group D_3

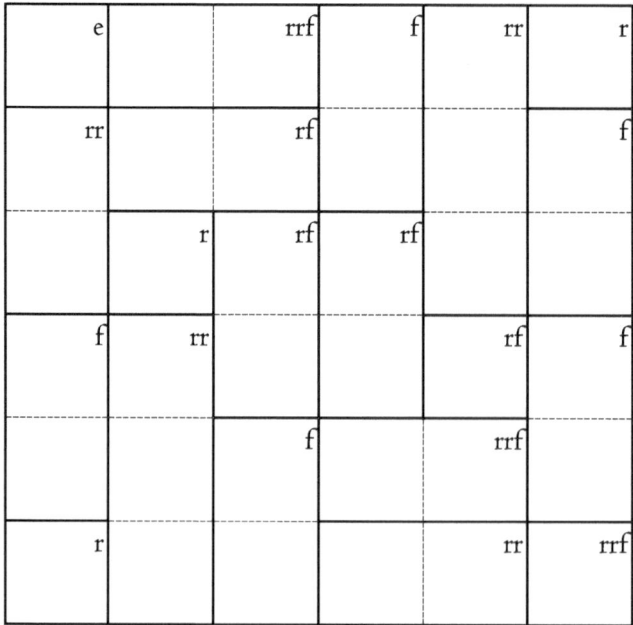

Fig. 4.11 Two puzzles over the dihedral group D_3

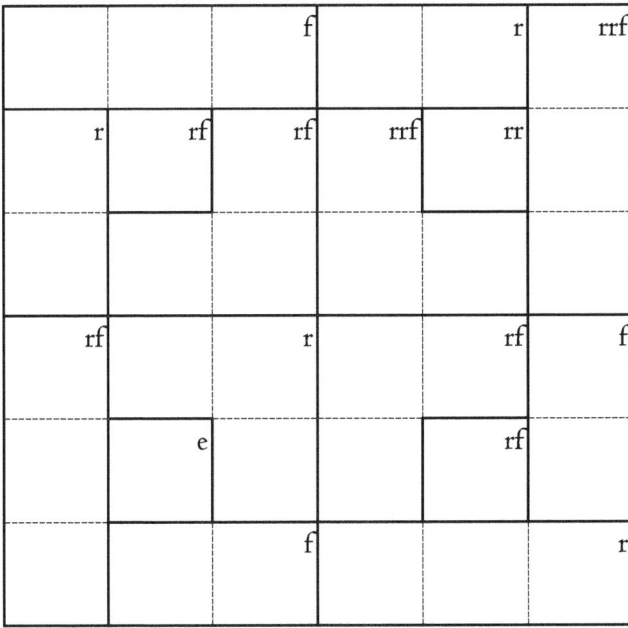

Fig. 4.12 Two puzzles over the dihedral group D_3

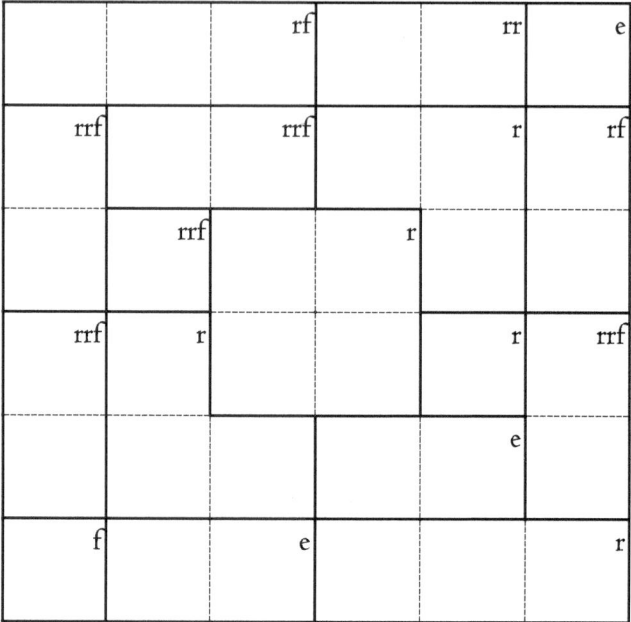

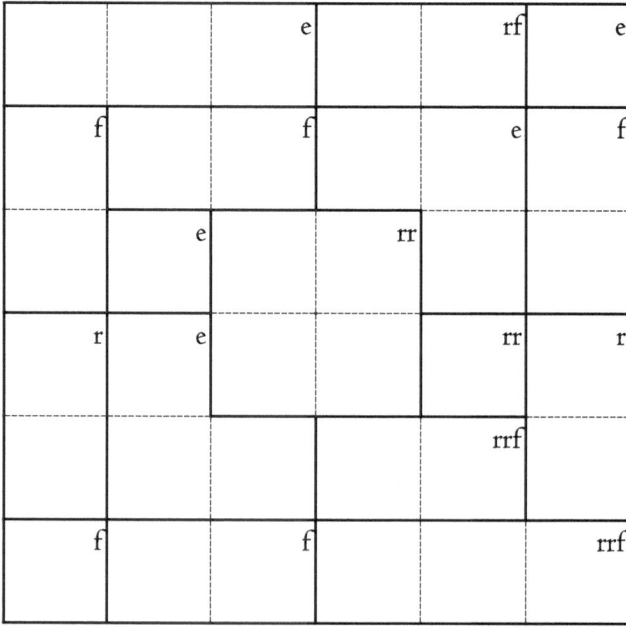

Fig. 4.13 Two puzzles over the dihedral group D_3

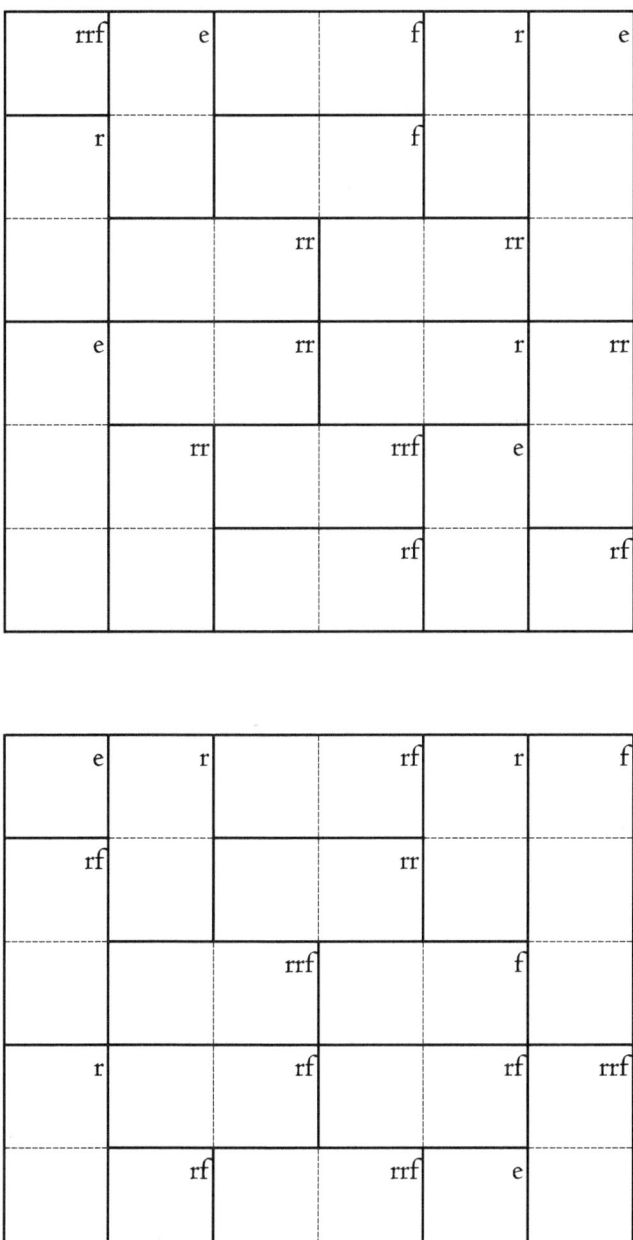

Fig. 4.14 Two puzzles over the dihedral group D_3

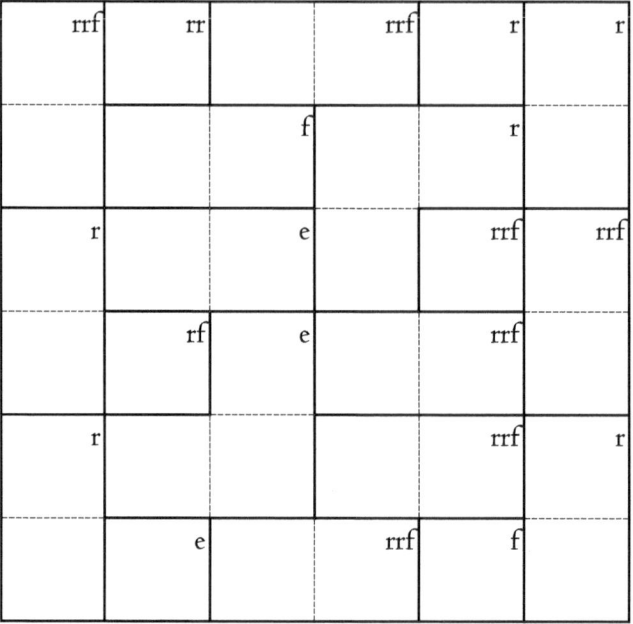

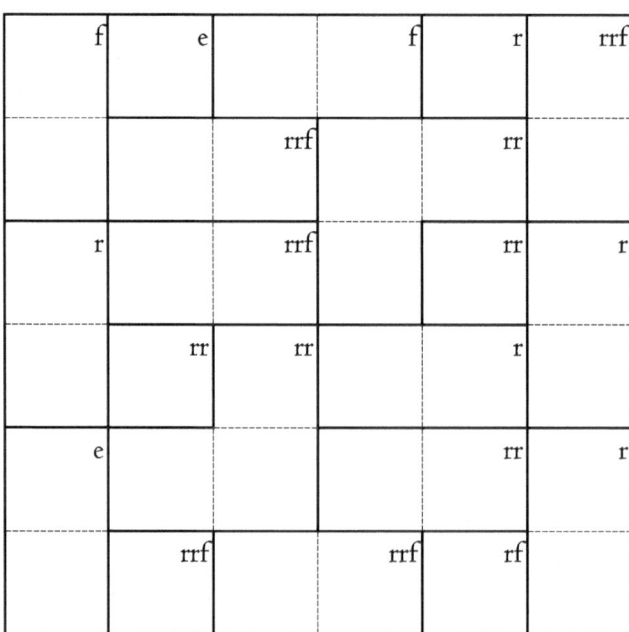

Fig. 4.15 Two puzzles over the dihedral group D_3

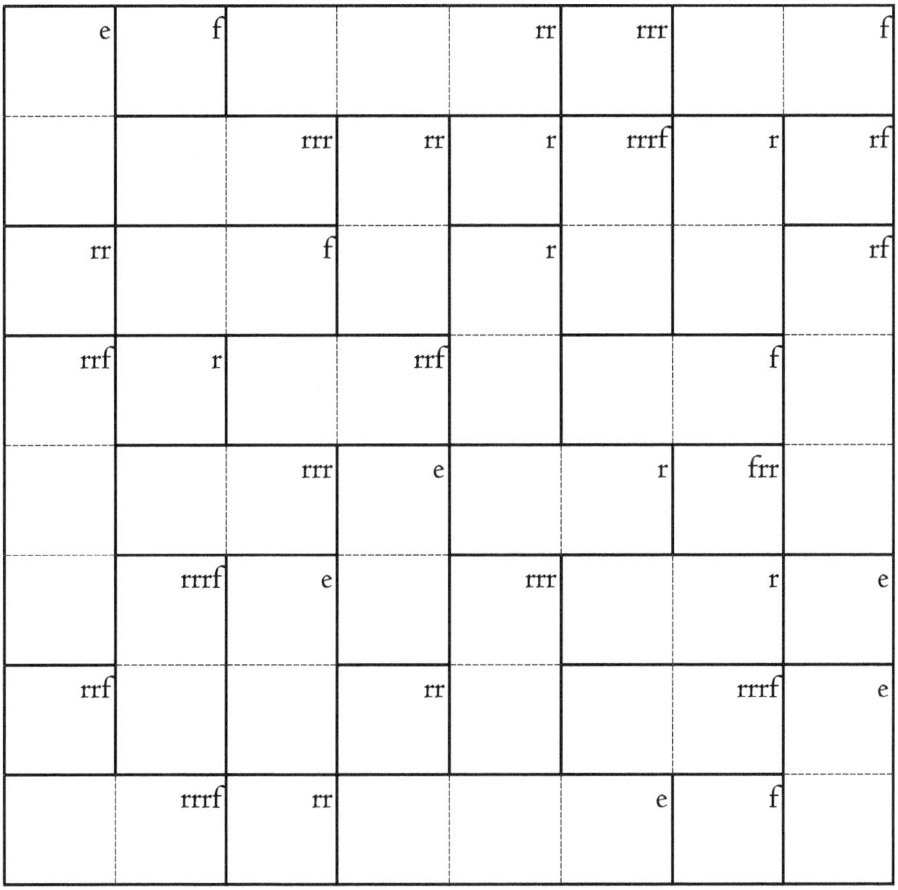

e	f			rr	rrr		f
		rrr	rr	r	rrrf	r	rf
rr		f		r			rf
rrf	r		rrf			f	
		rrr	e		r	frr	
	rrrf	e		rrr		r	e
rrf			rr			rrrf	e
	rrrf	rr			e	f	

Fig. 4.16 A puzzle over the dihedral group D_4

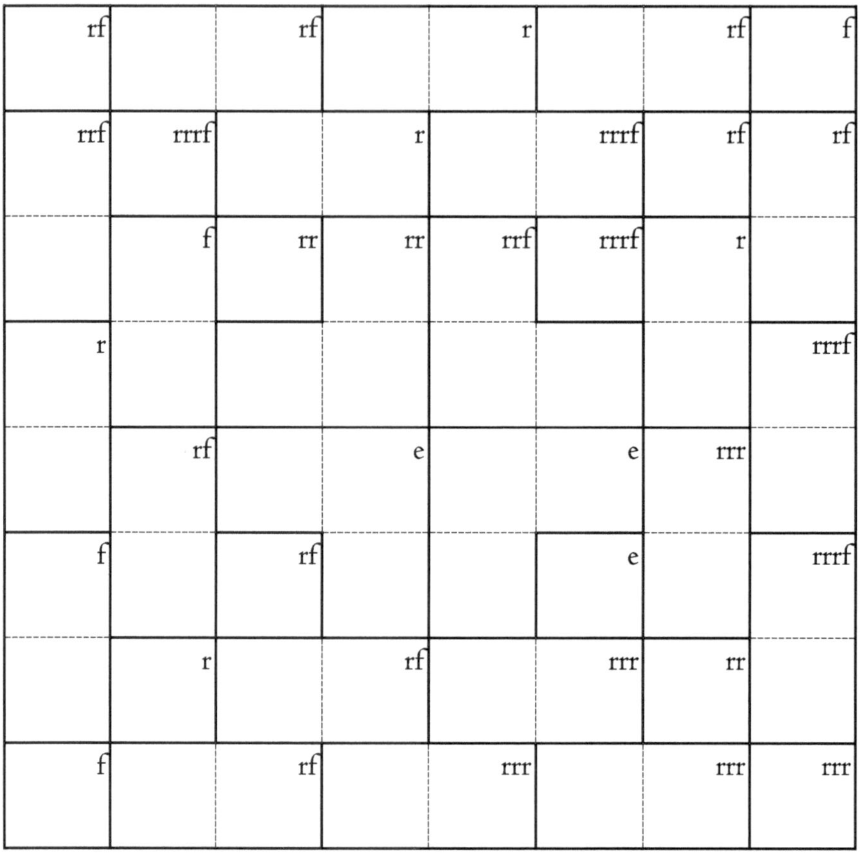

rf		rf		r		rf	f
rrf	rrrf		r		rrrf	rf	rf
	f	rr	rr	rrf	rrrf	r	
r							rrrf
	rf		e		e	rrr	
f		rf			e		rrrf
	r		rf		rrr	rr	
f		rf		rrr		rrr	rrr

Fig. 4.17 A puzzle over the dihedral group D_4

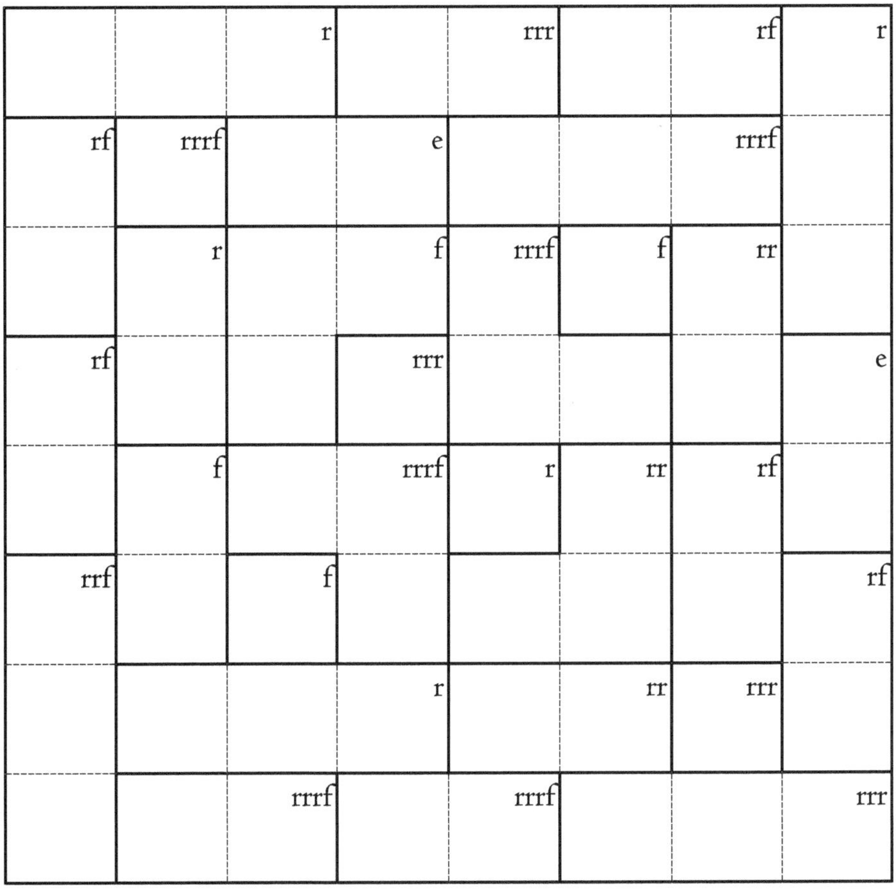

		r		rrr		rf	r
rf	rrrf		e			rrrf	
	r		f	rrrf	f	rr	
rf			rrr				e
	f		rrrf	r	rr	rf	
rrf		f					rf
			r		rr	rrr	
		rrrf		rrrf			rrr

Fig. 4.18 A puzzle over the dihedral group D_4

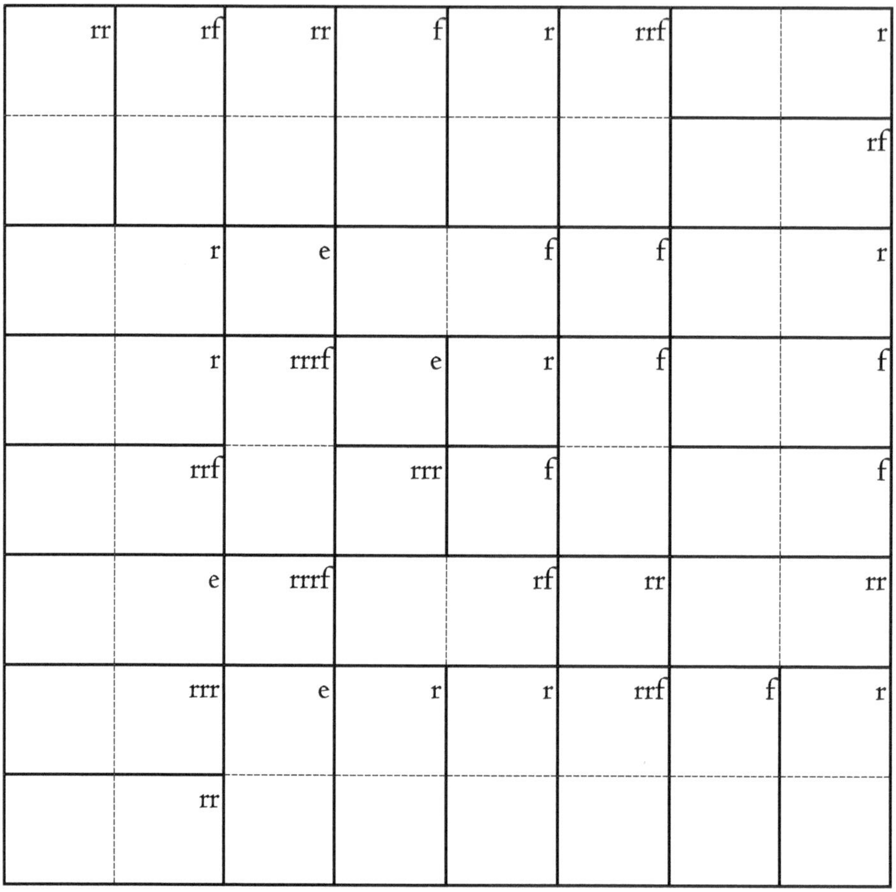

rr	rf	rr	f	r	rrf		r
							rf
	r	e		f	f		r
	r	rrrf	e	r	f		f
	rrf		rrr	f			f
	e	rrrf		rf	rr		rr
	rrr	e	r	r	rrf	f	r
	rr						

Fig. 4.19 A puzzle over the dihedral group D_4

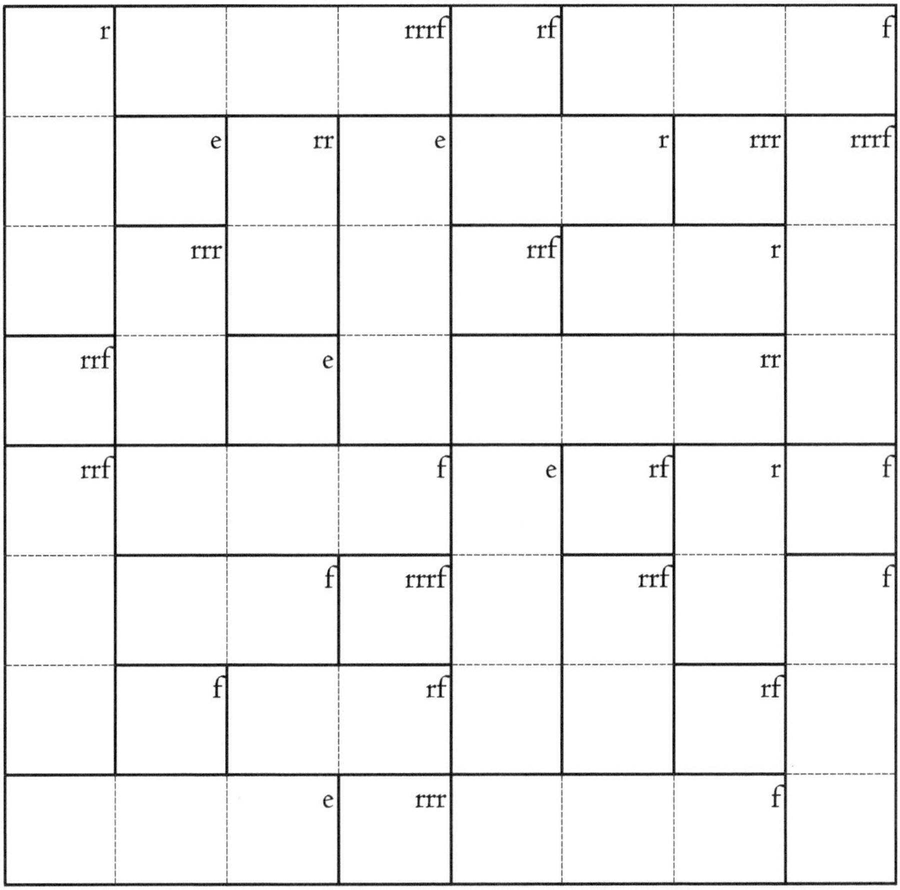

r			rrrf	rf			f
	e	rr	e		r	rrr	rrrf
	rrr			rrf		r	
rrf		e				rr	
rrf			f	e	rf	r	f
		f	rrrf		rrf		f
	f		rf			rf	
		e	rrr			f	

Fig. 4.20 A puzzle over the dihedral group D_4

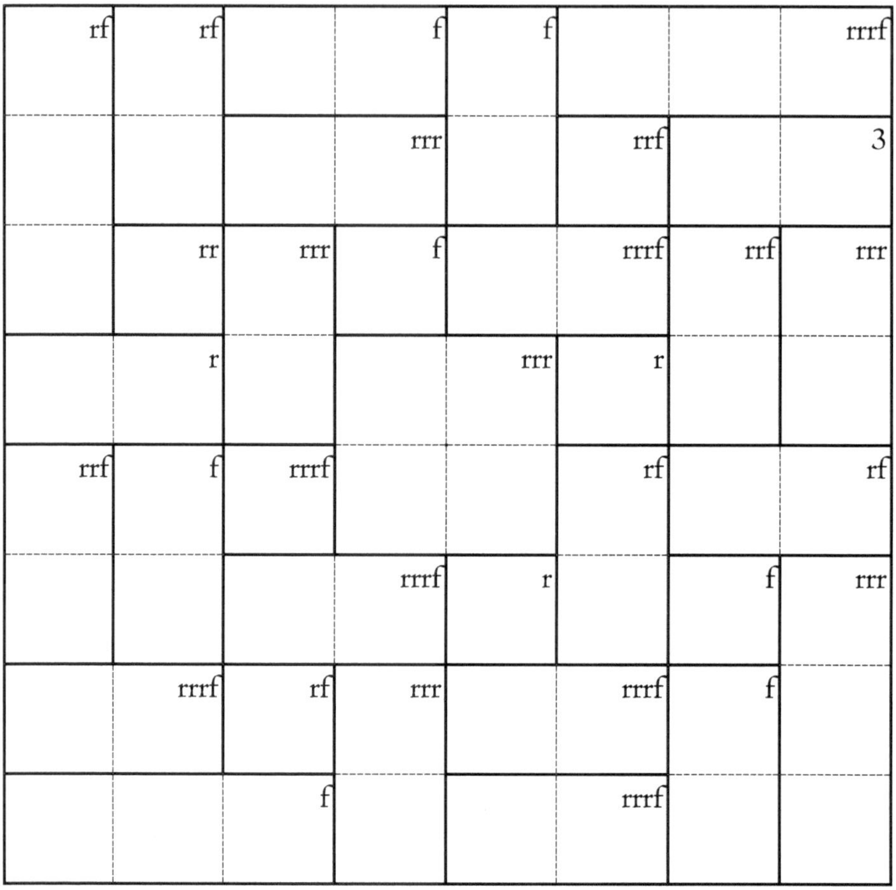

rf	rf		f	f			rrrf
			rrr		rrf		3
	rr	rrr	f		rrrf	rrf	rrr
	r			rrr	r		
rrf	f	rrrf			rf		rf
			rrrf	r		f	rrr
	rrrf	rf	rrr		rrrf	f	
		f			rrrf		

Fig. 4.21 A puzzle over the dihedral group D_4

4.2 Permutation Groups

Three Card Monte is a game where a dealer places three cards face down on a table, with exactly one being a Queen. The other player is briefly shown which card is the Queen, and then the dealer quickly shuffles all three cards by performing a sequence of swaps. The player wins if they can still identify the Queen after the shuffles; otherwise, the dealer wins. The game is commonly run by cardsharps who use sleight of hand to swap out the Queen and ensure they win. However, in this book, we insist that everyone plays fair.

Each possible move rearranges the cards in a way that each position gets sent to another, with a card being sent to every position and no two cards sent to the same position. A map between positions with these properties is called a *permutation*.

We want a method for naming each of the possible permutations in this game so we introduce what we call *cycle notation*. Here we use numbers inside parentheses to indicate all positions where cards are moved. When the number y directly follows x, this means that the card in position x gets moved to position y. If x is the last number in the cycle, the card in position x goes to the first position listed in the cycle.

Consider first the move that just swaps the card in the first position with the card in the second position, leaving the third one fixed. We can refer to this by saying "one goes to two, two goes to one" and write (12). The move that fixes the first card but swaps the others is referred to as (23), and the move that fixes only the card in the second position is referred to as (13).

We can shift each card so that the first card moves to the second position, the second card moves to the third position, and the final card moves to where the first card initially was. This permutation is denoted as (123). Reversing this move gives us the permutation (132). To represent no movement at all, we use the familiar letter e. These six permutations are depicted in Fig. 4.22.

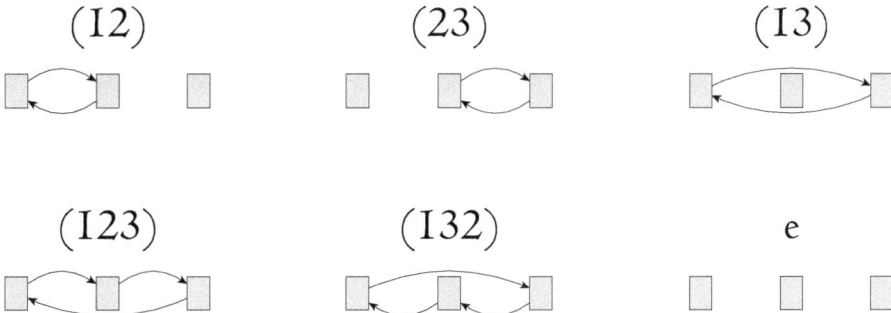

Fig. 4.22 The six basic moves in Three Card Monte

Notice that with this notation the symbols (123), (231), and (312) all represent the same permutation, and similarly, (132), (321) and (213) are also equivalent. Likewise, (12) = (21), (13) = (31), and (23) = (32). This should not be too difficult to bear since we are all used to number systems where there are multiple ways of writing the same number, such as the set of fractions, where $\frac{2}{4} = \frac{3}{6} = \frac{5}{10}$. Just as fractions have the concept of lowest terms, here each cycle can be written uniquely if we insist that the smallest element is listed first.

In a game of Three Card Monte, combinations of these six moves are executed rapidly in succession. Ultimately, every possible sequence of moves will be equivalent to a single one of the six permutations already discussed. For example, consider what happens when we first apply the permutation (12) and then apply the permutation (23). We can determine the resulting cycle by tracking where each card is moved during both operations.

The card in position one moves to position two in the first permutation, which to position three in the second. To determine the resulting cycle, we start with (13 and keep it open until we determine where the card originally in position three goes. The third card did not get moved at all in the first permutation, but did get moved to position two in the second, thus we can now write (132). There is only one position left for where the card in position two must have gone. By the process of elimination, it must go to position one. We can always double check by noting that the card is position two is sent to position one by (12), where it stays during (23). We can now close our parenthesis and be certain our cycle is (132).

We define a product on the collection of our six permutations, setting $a \times b$ to be the permutation we get from first applying b and then applying a. Applying permutations from right to left matches the notation for composition of functions found commonly in precalculus and calculus, as well as the notation we chose for dihedral groups. Our last calculation can now be written as $(23) \times (12) = (132)$.

It doesn't take long to construct a multiplication table for all possible products, and once that's done, we can see that this forms a group. The product is associative, being a composition of motions, and e serves as our identity element. The elements e, (12), (13), and (12) each act as their own inverses, while (123) and (132) are inverses of each other.

Before getting too excited about this brand new group, one might wish to examine how its table compares to the one for D_3. It turns out they are the same under a relabeling and reordering of the elements. First, relabel (123) as r, which means we must send $(123)^2 = (132)$ to r^2. Sending (12) to f determines the pairing between the rest of the elements in the two groups. For example (23) is $(12) \times (123)$ so it must go to $f \times r = r^2 f$. This leaves rf as the only possibility for (13) as the identity must go to the identity. Reordering our table after this relabeling leads to the identical table for D_3.

This does not mean that groups of permutations are not very interesting. Consider the extensions we would get from constructing the game of Four Card Monte, or the corresponding games for five cards or more. Taking the collection of all permutations for any of these cases gives us a finite group. Together, we call this collection the *symmetric groups*, and refer to the group on n cards as S_n.[5] For larger values of n, some permutations can no longer be written as a single cycle, but we can still write any permutation as a product of cycles, each with no numbers in common. For example, the permutation $(12)(34)$ in S_4 swaps both the first two and the last two cards, and cannot be represented as a single cycle $(x_1, x_2, \cdots, x_n)$.

For n greater than 3, these groups are no longer isomorphic to any dihedral group, though they are already large enough to make puzzle solving inconvenient. They are worth considering because they are not only important in their own right, but because subsets of large enough groups contain the complexities of entire subject of finite group theory. One of the many beautiful proofs in a first course in abstract algebra is the one that shows that every finite group, including all the ones we've seen so far, arises as some subset of a large enough permutation group.[6] In some sense, this means that the study of groups is a subset of the study of permutations.

In this book, we choose to include fewer puzzles over S_3 because they are equivalent to relabeled versions of the puzzles we recently considered for D_3. We still include some examples to give readers practice with cycle notation and to help them recognize connections between the two groups.

Puzzle Rules

Fill in the cells so that each element of the group S_3 appears exactly once in every row and column. The clues in each cage indicate the product of the cells when taken top to bottom or left to right for $n \times 1$ or $1 \times n$ shaped cages, and when taken clockwise for L-shaped cages. The following tables (Figs. 4.23, 4.24, and 4.25) may be helpful for solving these puzzles (Figs. 4.26, 4.27, and 4.28).

---

[5] Symmetric groups played a crucial role in the development of group theory, and were studied by Lagrange in his study of roots of polynomials of fourth degree, long before Abel and Galois discovered and dug deeper to the theory of groups in their search for solutions to fifth degree equations [8].

[6] This result is known as Cayley's theorem, named after the mathematician Arthur Cayley, In addition to this result, Cayley is credited as the first mathematician to take an axiomatic, or rule based, approach to groups. This approach is used here in this book and is the standard approach given in textbooks today [8, 12, 18].

Fig. 4.23 The group operation table for the group S_3

×	e	(123)	(132)	(12)	(13)	(23)
e	e	(123)	(132)	(12)	(13)	(23)
(123)	(123)	(132)	e	(13)	(23)	(12)
(132)	(132)	e	(123)	(23)	(12)	(13)
(12)	(12)	(23)	(13)	e	(132)	(123)
(13)	(13)	(12)	(23)	(123)	e	(132)
(23)	(23)	(13)	(12)	(132)	(123)	e

e	(123)	(132)	(12)	(13)	(23)
$(123) \cdot (132)$	$e \cdot (123)$	$e \cdot (132)$	$e \cdot (12)$	$e \cdot (13)$	$e \cdot (23)$
$(132) \cdot (123)$	$(123) \cdot e$	$(132) \cdot e$	$(123) \cdot (23)$	$(123) \cdot (12)$	$(123) \cdot (13)$
	$(12) \cdot (23)$	$(12) \cdot (13)$	$(132) \cdot (13)$	$(132) \cdot (23)$	$(132) \cdot (12)$
	$(13) \cdot (12)$	$(13) \cdot (23)$	$(12) \cdot e$	$(12) \cdot (132)$	$(12) \cdot (123)$
	$(23) \cdot (13)$	$(23) \cdot (12)$	$(13) \cdot (123)$	$(13) \cdot e$	$(13) \cdot (132)$
			$(23) \cdot (132)$	$(23) \cdot (123)$	$(23) \cdot e$

Fig. 4.24 All possibilities for two-cell cages in S_3

Fig. 4.25 Inverses of elements in S_3

a	e	(123)	(132)	(12)	(13)	(23)
a^{-1}	e	(132)	(123)	(12)	(13)	(23)

		e		(13)	e
(13)		(23)	(13)	(12)	e
	(13)		(23)		
(12)	e		(132)	(123)	(123)
		e		(12)	
(123)		(13)			(123)

(23)		(23)	(23)	e	(12)
(123)		(12)			(132)
	(12)	(13)	(23)		
(132)	(13)			(123)	(132)
		(12)		(123)	
(132)				(132)	(123)

Fig. 4.26 Two puzzles over the symmetric group S_3

		(132)		e	(123)
(23)	(12)	(23)	(123)	(132)	
e		e		(123)	(132)
	e			(23)	
		e			(123)

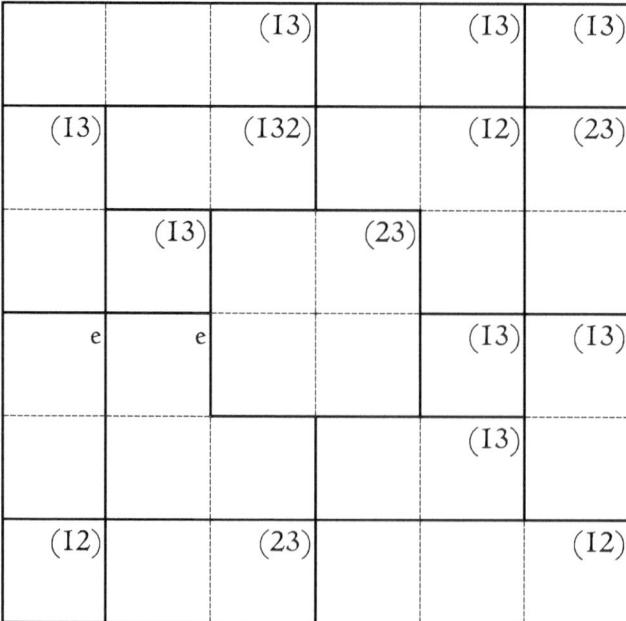

Fig. 4.27 Two puzzles over the symmetric group S_3

e	(12)		(123)	(13)	(12)
(23)			(123)		
		(13)		(132)	
e		(23)		(13)	(13)
	e		(123)	(132)	
			(12)		(23)

e	(23)		(23)	e	(23)
		(12)		(23)	
(13)		(12)		(132)	(23)
	(12)	(123)		(12)	
(123)				(12)	(23)
	(13)		(123)	(12)	

Fig. 4.28 Two puzzles over the symmetric group S_3

4.3   The Quaternion Group

In this book, we wish to introduce all groups of size eight or less up to isomorphism. By this, we mean that every possible group of order eight or less will be isomorphic to a group from this book. We are one group away from this goal, and we will now construct it in this section. This new group is a little more involved, and is usually created by first constructing a larger mathematical system, and then picking selected elements. Here, we take a slightly different approach.

The integers 1 and -1 form a group under multiplication which is isomorphic to $\mathbb{Z}_2$, as shown in Fig. 4.29. Here, both elements squared are equal to 1, and no element squared equals -1. Let's change that.

For any subset of integers, the squares cannot be negative, but what would happen if we just forced a square root of negative one? We create a new element whose square is -1, give it a name like i, and then $i^2 = -1$, just like that! If we try adding i into our group, what would happen? How many other elements would we need to add? Looking at Fig. 4.30 we see we need elements to equal $(-1) \times i$ and $i \times (-1)$. It makes the most sense just to add a single element we call $-i$ for this purpose. Would we need to add anything else?

Using the fact that $-i = (-1) \times i = i \times (-1)$ together with associativity, we find that

$$(i)(-i) = (i)(i(-1)) = (ii)(-1) = (-1)^2 = 1,$$

$$(-i)(i) = ((-1)i)(i) = (-1)(ii) = (-1)^2 = 1,$$

and

$$(-i)(-i) = ((-1)i)(-i) = (-1)(i(-i)) = (-1)(1) = (-1).$$

Now that we've found all possible products, we arrive at the table shown in Fig. 4.31. This is indeed a group with identity 1, where 1 and -1 are their own inverses and i and $-i$ are inverses of each other. We chose to list the elements in the order $1, i, -1, -i$ to make it easy to see this is isomorphic to the group $\mathbb{Z}_4$.

Fig. 4.29 The multiplicative group on $\{-1, 1\}$

$\times$	1	-1
1	1	-1
-1	-1	1

Fig. 4.30 The multiplicative group on $\{-1, 1\}$

$\times$	1	-1	i
1	1	-1	i
-1	-1	1	
i	i		-1

Fig. 4.31 The cyclic group generated by i

×	1	i	-1	$-i$
1	1	i	-1	$-i$
i	i	-1	$-i$	1
-1	-1	$-i$	1	i
$-i$	$-i$	1	i	-1

Fig. 4.32 Forming a new group on the set $\{1, -1, i, -i, j, -j, k, -k\}$

×	1	-1	i	$-i$	j	$-j$	k	$-k$
1	1	-1	i	$-i$	j	$-j$	k	$-k$
-1	-1	1	$-i$	i				
i	i	$-i$	-1	1				
$-i$	$-i$	i	1	-1				
j	j				-1			
$-j$	$-j$					-1		
k	k						-1	
$-k$	$-k$							-1

Notice that when $-i$ is squared, we get the same value as when i is squared. The act of adding a root of -1 actually leads to adding a pair of roots, with each one being the negative of the other. Both roots behave identically. In fact, swapping all occurrences of i and $-i$ in the table, and then reordering leads to the same exact table we've already found.

Now we're ready to search for the final group of order eight, and the idea that works for this is to try to make a multiplicative group out of roots of -1. We already know that roots come in pairs, and that we must include both -1 and 1, so this leaves room for four more roots. Rolling forward along the alphabet from i, lets call these $j, -j, k$, and $-k$.

Our goal here is only to ask what a group on these eight distinct elements would look like if it exists, assuming that -1 behaves as we would normally expect, by mapping $1, i, j$ and k to $-1, -i, -j$ and $-k$, and mapping $-1, -i, -j$ and $-k$ back to $1, i, j$ and k. Is this enough to determine an entire eight-by-eight table as shown so far in Fig. 4.32?

For -1 to behave as expected with respect to all our group elements, we need the following rules, which we will use in different ways by both "bringing in" and "pulling out" negatives.

$$(-1)i = i(-1) = -i$$

$$(-1)(-i) = (-i)(-1) = i$$

$$(-1)j = j(-1) = -j$$

$$(-1)(-j) = (-j)(-1) = j$$

$$(-1)k = k(-1) = -k$$

$$(-1)(-k) = (-k)(-1) = k$$

$$(-1)(-1) = 1$$

The inverses of our six roots of negative one are now determined by associativity together with the rules for negatives above. For one example,

$$(-j)j = ((-1)j)(j) = (-1)(jj) = (-1)(-1) = 1$$

and

$$j(-j) = (j)(j(-1)) = (jj)(-1) = (-1)(-1) = 1,$$

showing that j and $-j$ are inverses of each other. Similar reasoning shows that both i and $-i$ and k and $-k$ are pairs of inverses of each other.

We now know how 1 and -1 interact with everything, and how $i, -i, j, -j, k,$ and $-k$ interact with themselves and their negatives. We can already complete a group operation table up to what is shown in Fig. 4.33. The only things left to determine are the products of the forms $\pm i \times \pm j, \pm j \times \pm k,$ and $\pm k \times \pm i$.

Let's try to figure out what i times j needs to be to make things work. We went into this assuming that all eight elements of our group were distinct. If we chose to define $i \times j = j$ or $i \times j = -j$ we could multiply on the right by $-j$ to conclude $i = 1$ or $i = -1$, thus contradicting our assumption of distinctness. Similarly, $i \times j$

Fig. 4.33 Progress constructing a new group on the set $\{1, -1, i, -i, j, -j, k, -k\}$

$\times$	1	-1	i	$-i$	j	$-j$	k	$-k$
1	1	-1	i	$-i$	j	$-j$	k	$-k$
-1	-1	1	$-i$	i	$-j$	j	$-k$	k
i	i	$-i$	-1	1				
$-i$	$-i$	i	1	-1				
j	j	$-j$			-1	1		
$-j$	$-j$	j			1	-1		
k	k	$-k$					-1	1
$-k$	$-k$	k					1	-1

cannot be i or $-i$, otherwise multiplying on the left by $-i$ would reveal j is equal to 1 or -1, which is again a contradiction.

We cannot set $i \times j$ to 1 either, because then

$$ij = i(-i)$$

as both equal 1. Multiplying both sides by $(-i)$ on the left would cause j to equal $(-i)$.[7] Similarly, $i \times j$ cannot be -1, otherwise

$$ij = i^2$$

and multiplication on the left by $(-i)$ again causes trouble.

There are only two possibilities left among our eight elements for $i \times j$. It must be either k or $-k$. We will choose k, but oddly enough, this last choice doesn't matter. Both options actually work and if we had chosen $-k$, we would have gotten an equivalent group where k and $-k$ are simply interchanged in all instances.

Once we have decided that $i \times j = k$ we can manipulate this equation to get all of the remaining twenty-four possible products. Multiplying the equation

$$i \times j = k$$

on the left by $-i$ and on the right by k gives us

$$j \times k = i.$$

Multiplying this new equation on the left by $-j$ and on the right by i gives us

$$k \times i = j.$$

Now if we take this last equation $k \times i = j$ and multiply on the right by i, we can conclude $j \times i = -k$. Similar reasoning gives us $k \times j = -i$ and $i \times k = -j$. This shows that the group we are making is not abelian. We can summarize these rules with a picture such as Fig. 4.34, which reveals the product of two elements in order when following the arrows, and the negative of the product when going against them. This circle, together with our previous results regarding negatives and roots of -1, is really all we need to find every possible product. This makes this group particularly easy to navigate without having to resort to looking at its table, after a little practice.

The table we get from this construction is given at the end of this section, and forms a group that we call Q, or the *quaternion group*.[8] With this, we have now

---

[7] This is also an immediate consequence of the fact that in a group, inverses are unique and we already know that the inverse of i is $-i$. Since we did not prove this fact, we include the argument above.

[8] Though the existence of identity and inverses can be easily seen from the table, and though we used associativity to find some of the products, the fact that the entire group is associative is less

Fig. 4.34 A diagram
summarizing the
multiplication in Q

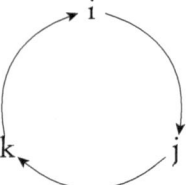

produced the last of the five non-isomorphic groups of order eight, with the others
being $\mathbb{Z}_8$, $\triangle(\{1, 2, 3\})$, $U(16)$, and D_4.[9] This being done, we now introduce our last
challenging collection of puzzles.

Puzzle Rules

Fill in the cells so that each element of the group Q appears exactly once in every
row and column. The clues in each cage indicate the product of the cells when
taken top to bottom or left to right for $n \times 1$ or $1 \times n$ shaped cages, and when taken
clockwise for L-shaped cages. The following tables (Figs. 4.35, 4.36, and 4.37) may
be helpful for solving these puzzles (Figs. 4.38, 4.39, 4.40, 4.41, 4.42, and 4.43).

Fig. 4.35 The group
operation table for the
quaternion group Q

$\times$	1	-1	i	$-i$	j	$-j$	k	$-k$
1	1	-1	i	$-i$	j	$-j$	k	$-k$
-1	-1	1	$-i$	i	$-j$	j	$-k$	k
i	i	$-i$	-1	1	k	$-k$	$-j$	j
$-i$	$-i$	i	1	-1	$-k$	k	j	$-j$
j	j	$-j$	$-k$	k	-1	1	i	$-i$
$-j$	$-j$	j	k	$-k$	1	-1	$-i$	i
k	k	$-k$	j	$-j$	$-i$	i	-1	1
$-k$	$-k$	k	$-j$	j	i	$-i$	1	-1

obvious. The easiest way to show associativity is perhaps to show that this group sits inside a
larger structure that is more easily proven associative. That structure is the ring (or skew-field)
of the quaternions. This number system has both a rich and beautiful history and an incredible
number of diverse applications [3, 9, 21]. Indeed, the quaternion group is usually constructed by
picking these eight elements out of this bigger structure, but we chose to take a faster approach.

[9] The proof that these are the only ones is not included and requires more theory than we've
developed here.

	1	-1	i	-i	j	-j	k	-k
$i\cdot -i$		$1\cdot -1$	$1\cdot i$	$1\cdot -i$	$1\cdot j$	$1\cdot -j$	$1\cdot k$	$1\cdot -k$
$-i\cdot i$		$-1\cdot 1$	$-1\cdot -i$	$-1\cdot i$	$-1\cdot -j$	$-1\cdot j$	$-1\cdot -k$	$-1\cdot k$
$j\cdot -j$			$i\cdot 1$	$i\cdot -1$	$i\cdot -k$	$i\cdot k$	$i\cdot j$	$i\cdot -j$
$-j\cdot j$			$-i\cdot -1$	$-i\cdot 1$	$-i\cdot k$	$-i\cdot -k$	$-i\cdot -j$	$-i\cdot j$
$k\cdot -k$			$j\cdot k$	$j\cdot -k$	$j\cdot 1$	$j\cdot -1$	$j\cdot -i$	$j\cdot i$
$-k\cdot k$			$-j\cdot -k$	$-j\cdot k$	$-j\cdot -1$	$-j\cdot 1$	$-j\cdot i$	$-j\cdot -i$
			$k\cdot -j$	$k\cdot j$	$k\cdot i$	$k\cdot -i$	$k\cdot 1$	$k\cdot -1$
			$-k\cdot j$	$-k\cdot -j$	$-k\cdot -i$	$-k\cdot i$	$-k\cdot -1$	$-k\cdot 1$

Fig. 4.36 All possibilities for two cell cages in Q

Fig. 4.37 Inverses of elements in Q

a	1	-1	i	$-i$	j	$-j$	k	$-k$
a^{-1}	1	-1	$-i$	i	$-j$	j	$-k$	k

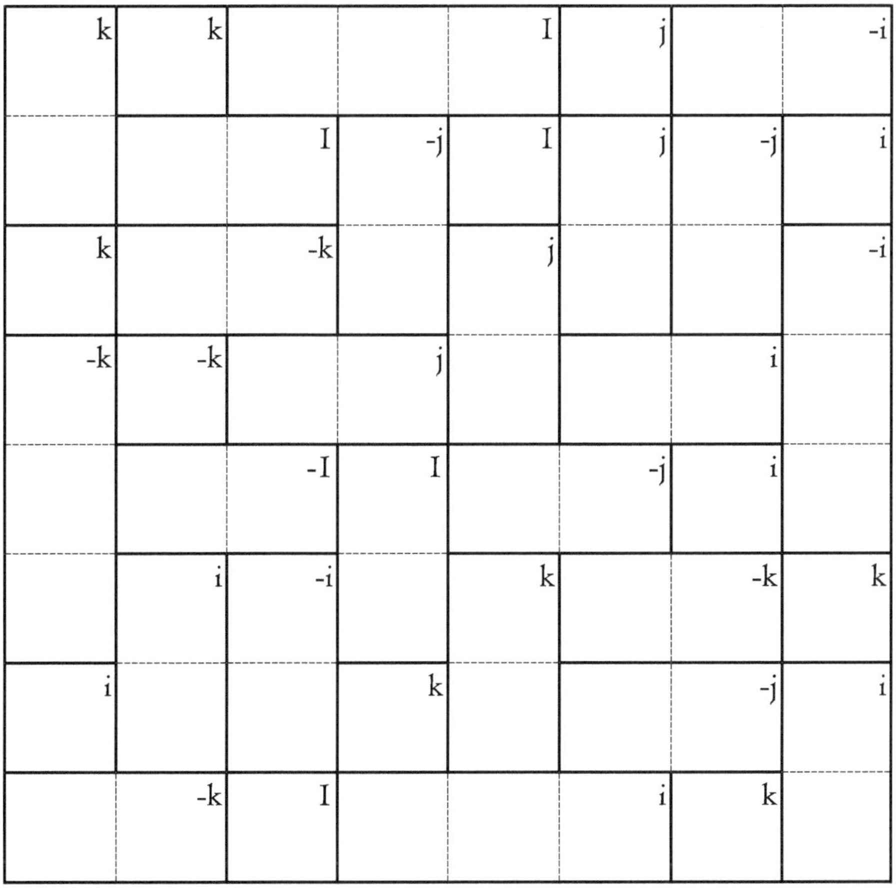

k	k			I	j		-i
		I	-j	I	j	-j	i
k		-k		j			-i
-k	-k		j			i	
		-I	I		-j	i	
	i	-i		k		-k	k
i			k			-j	i
	-k	I			i	k	

Fig. 4.38 A puzzle over the quaternion group Q

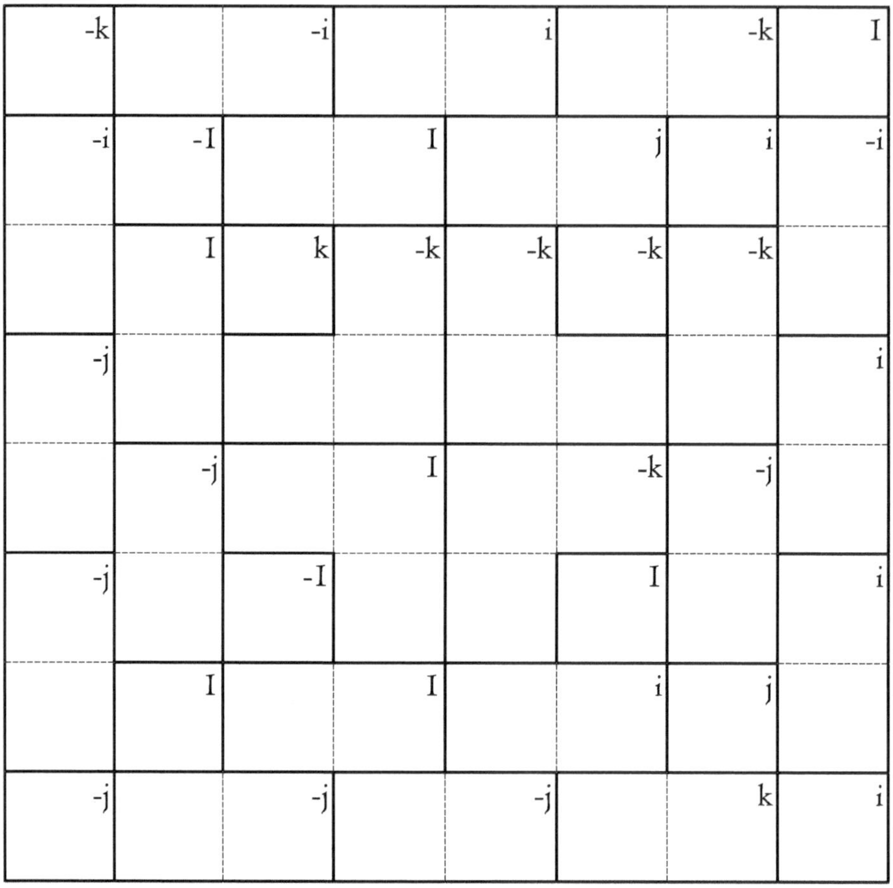

-k		-i		i		-k	I
-i	-I		I		j	i	-i
	I	k	-k	-k	-k	-k	
-j							i
	-j		I		-k	-j	
-j		-I			I		i
	I		I		i	j	
-j		-j		-j		k	i

Fig. 4.39 A puzzle over the quaternion group Q

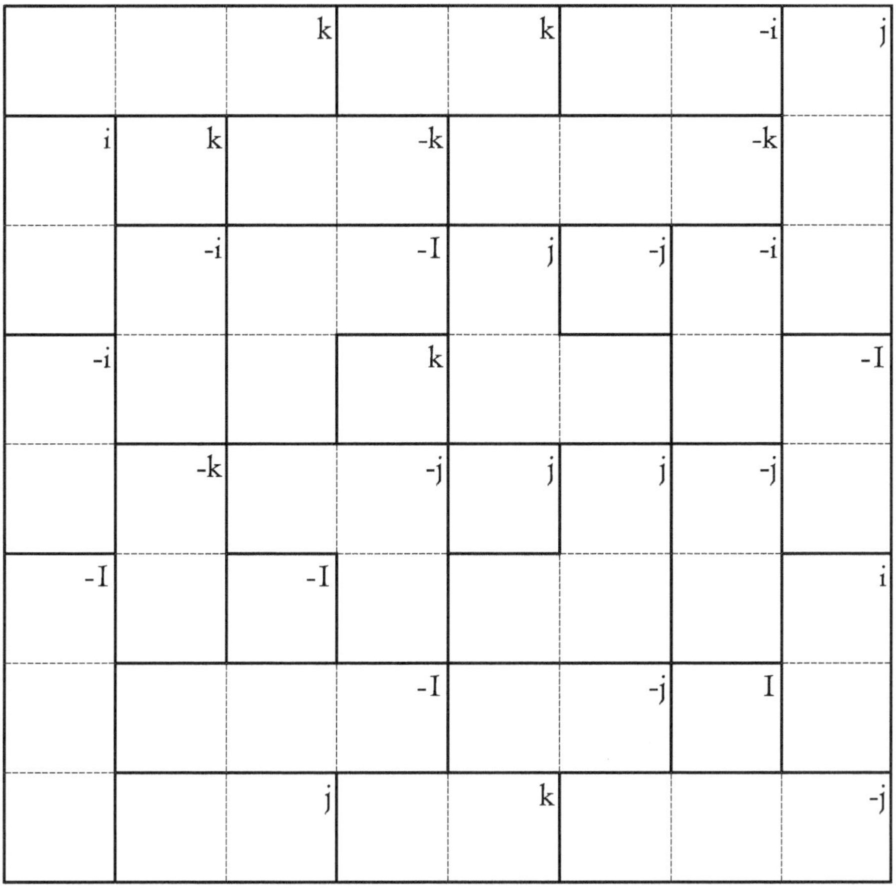

		k		k		-i	j
i	k		-k			-k	
	-i		-I	j	-j	-i	
-i			k				-I
	-k		-j	j	j	-j	
-I		-I					i
			-I		-j	I	
		j		k			-j

Fig. 4.40 A puzzle over the quaternion group Q

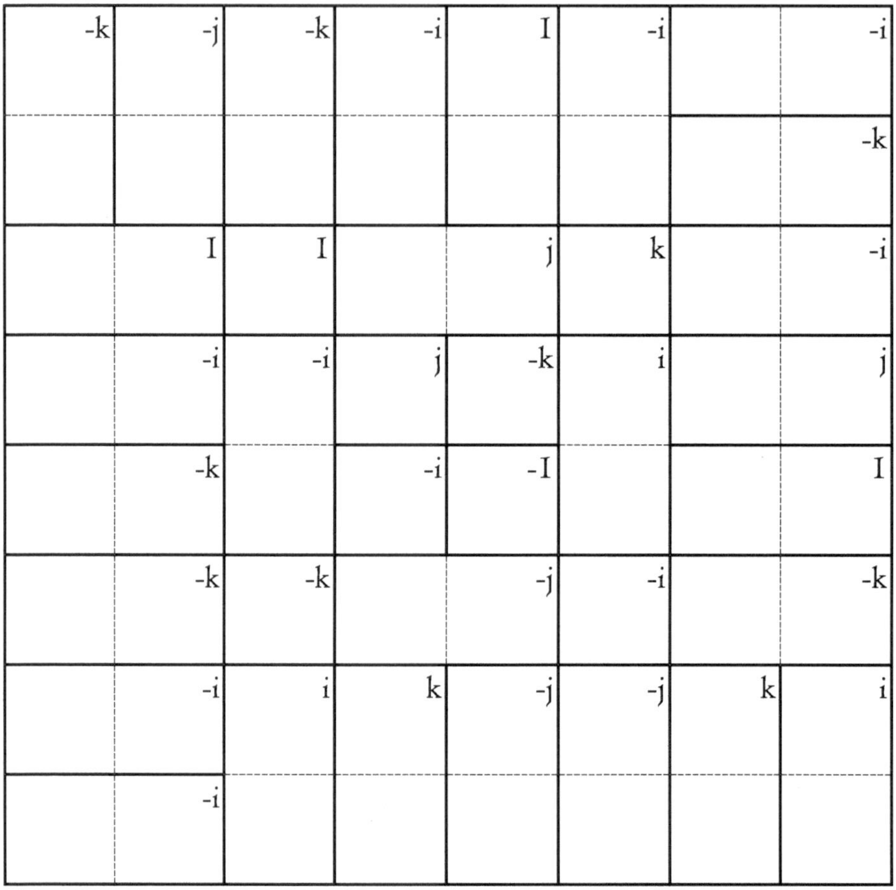

-k	-j	-k	-i	I	-i		-i
							-k
	I	I		j	k		-i
	-i	-i	j	-k	i		j
	-k		-i	-I			I
	-k	-k		-j	-i		-k
	-i	i	k	-j	-j	k	i
	-i						

Fig. 4.41 A puzzle over the quaternion group Q

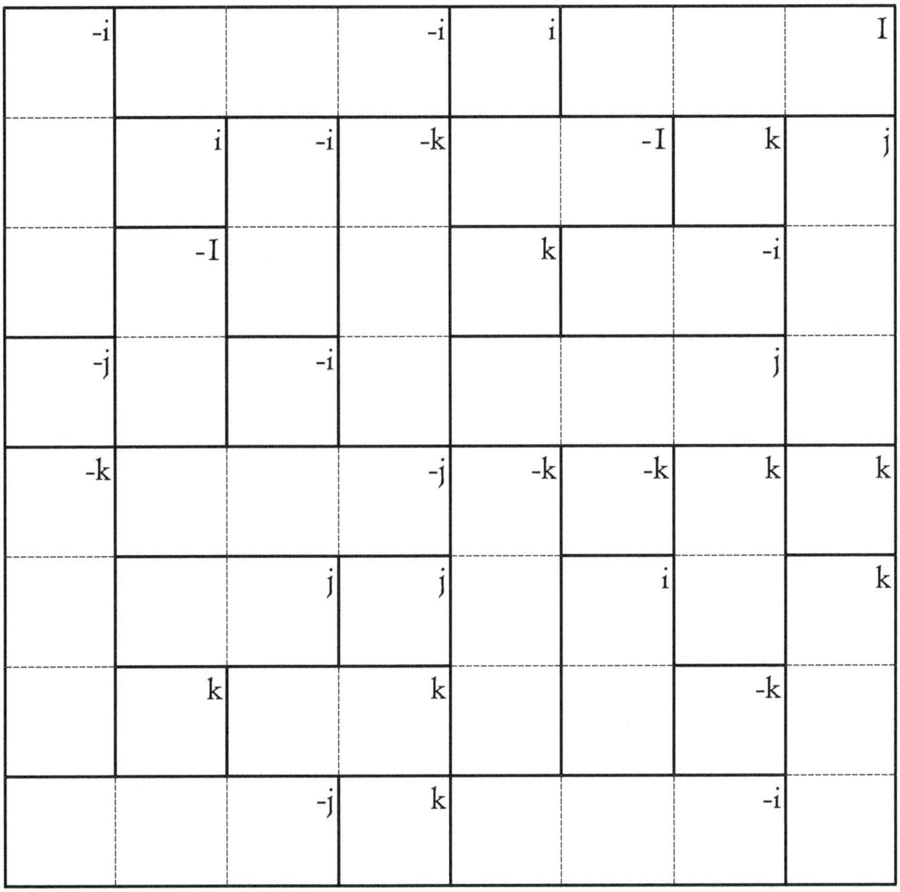

-i			-i	i			I
	i	-i	-k		-I	k	j
	-I			k		-i	
-j		-i				j	
-k			-j	-k	-k	k	k
		j	j		i		k
	k		k			-k	
		-j	k			-i	

Fig. 4.42 A puzzle over the quaternion group Q

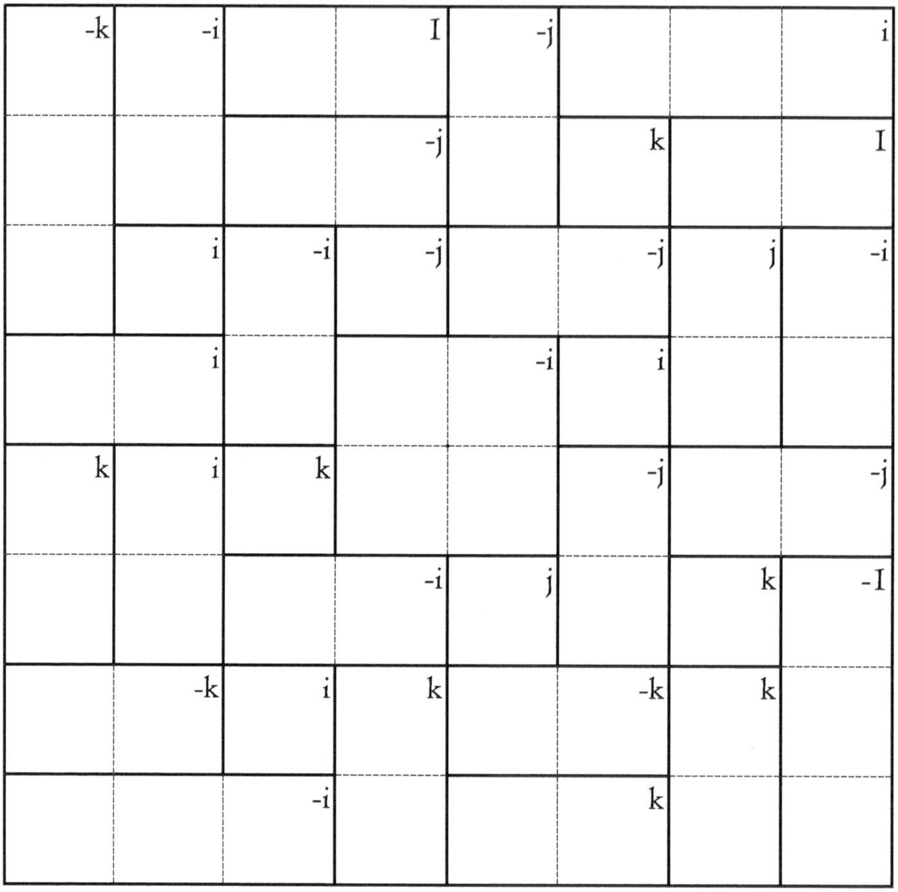

-k	-i		I	-j			i
			-j		k		I
	i	-i	-j		-j	j	-i
	i			-i	i		
k	i	k			-j		-j
			-i	j		k	-I
	-k	i	k		-k	k	
		-i			k		

Fig. 4.43 A puzzle over the quaternion group Q

Further Tricks and Techniques

5.1   Latin Square Tricks and the Group Product of All Elements

What happens if we take the product of all the entries in a single row or column of a puzzle over an abelian group? The cells must contain each element of the group exactly once, and since the group is abelian, the order of multiplication does not matter. Let's first consider a small example.

In the case of $\mathbb{Z}_7$, the elements are $\{0, 1, 2, 3, 4, 5, 6\}$. To get the group product, which here means the sum, of any row or column we would need to find $0 + 1 + 2 + 3 + 4 + 5 + 6$. We could just add these up, but note that this sum is equal to the number of squares shown in Fig. 5.1, which we will call S.

Instead of counting the squares directly, we could find S in another way.[1] If we take two such triangles of squares, then $2S$ is the number of squares in the diagram shown in Fig. 5.2. If we slide those squares together, we get a 6 by 7 rectangle, which must therefore contain 42 squares. Since $2 \times S = 42$ we have $S = 21$.

The advantage to this method is that it can be adjusted to work for any other sums as well. If we let S be the sum $0 + 1 + 2 + \cdots + (n - 1)$, we can put two triangles of height $n - 1$ together, and use the same trick to get the equation

$$2S = (n - 1)n.$$

This shows that

$$S = 0 + 1 + 2 + \cdots + (n - 1) = \frac{1}{2}(n - 1)n$$

and provides us with a formula for the sum of all the elements in the group $\mathbb{Z}_n$.

---

[1] This method is often incorrectly attributed to Gauss partly due to a story popularized in E. T. Bell's controversial book "Men of Mathematics." [4] There is no actual evidence for this story, and Brian Hayes has an excellent article on the history and spread of the legendary tale [10].

© The Author(s), under exclusive license to Springer Nature Switzerland AG 2024   121
D. Nacin, *The Group Theory Puzzle Book*, SUMS Readings,
https://doi.org/10.1007/978-3-031-70091-0_5

Fig. 5.1 A total
$0 + 1 + 2 + 3 + 4 + 5 + 6$ of
squares

Fig. 5.2 Two pyramids of
squares

Fig. 5.3 A puzzle over $\mathbb{Z}_4$

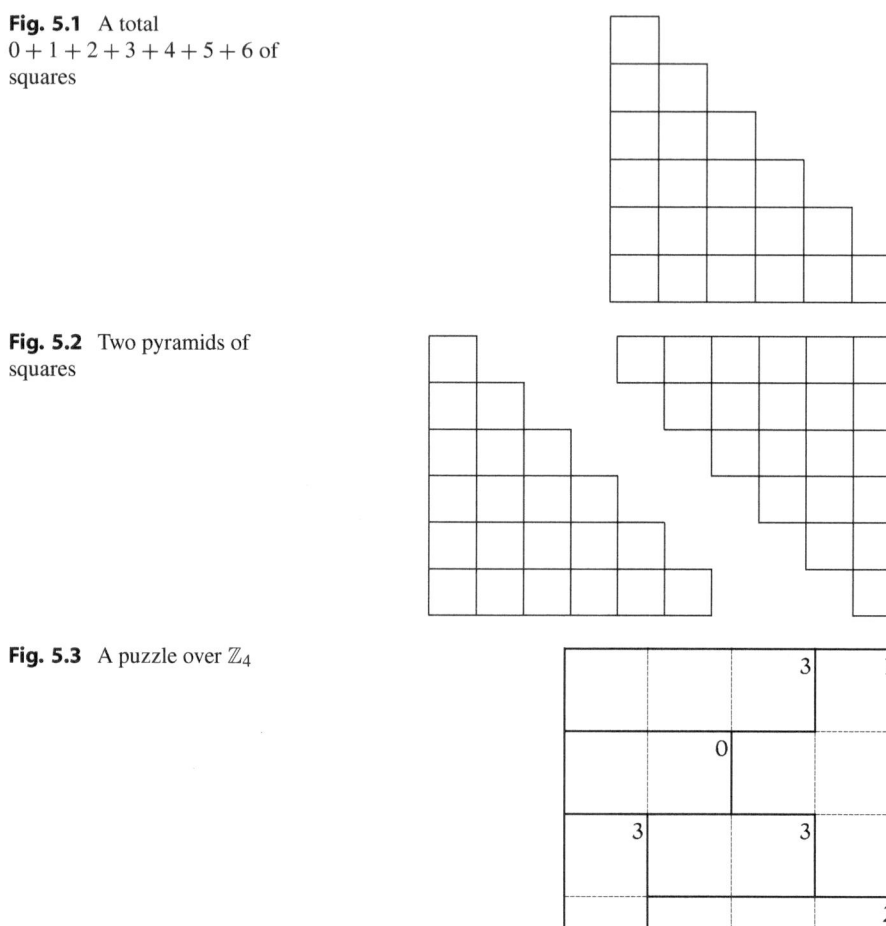

If n is odd, then $\frac{1}{2}(n - 1)$ is an integer, and this product is $\left(\frac{1}{2}(n - 1)\right)n$ and thus a multiple of n. In arithmetic modulo n, this is simply zero.

If n is even, then this product is $\frac{n}{2}(n - 1)$, which equals $-\frac{n}{2}$ in $\mathbb{Z}_n$, as $n - 1$ equals -1. However, since

$$-\frac{n}{2} = n + -\frac{n}{2} = \frac{n}{2},$$

we can conclude that in this case, the sum of the entries in any row or column is always exactly half of n.

For a first example of how we might use this, consider the puzzle shown in Fig. 5.3 over $\mathbb{Z}_4$. We know each row must add to half of 4, which is 2. In the top row, the first three cells add to 3, so the last cell must contain the solution to $3 + x = 2$.

Fig. 5.4 A partially solved
puzzle over $\mathbb{Z}_4$

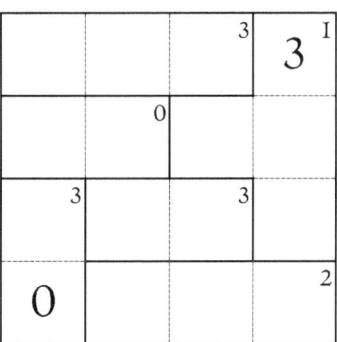

Thus this last cell contains the number 3. The bottom row must also add to 2. Since
the three rightmost cells there add to 2, the last cell must satisfy $2 + x = 2$. Thus
the leftmost cell there must contain 0. The two cells we just revealed are in two-cell
cages, so this technique has put us well on the way to solving our puzzle, as shown
in Fig. 5.4.

What about abelian groups that are not of the form $\mathbb{Z}_n$ for some n? How do we
find the product of all elements here? Well it can help to divide all elements into two
categories, placing the ones that are their own inverse into one set I, and ones that
are not into another set N. The order does not matter since our group is abelian, so if
we multiply all the elements in N, we can write each element next to its inverse and
allow them to cancel each other out. This reveals that the product of all elements in
N is just the identity. Therefore the product of all of the elements in the group is
equal to the product of the elements in I.

For an example, consider the group $U(16)$. Here, $I = \{1, -1, 7, -7\}$ and $N =$
$\{3, -3, 5, -5\}$. To get the product of all the elements in I we can use the fact that
$7 \times (-7) = -1$, so the product of all four is $(-1)^2$, or 1. Thus for puzzles over
$U(16)$ we know the product of any row or column is always 1.

For the symmetric difference groups, every element is its own inverse, so our
trick of just multiplying the elements in I will not save us any time. We can still try
to group our elements carefully before multiplying them together. As

$$\{1\}\Delta\{2, 3\} = \{2\}\Delta\{1, 3\} = \{3\}\Delta\{1, 2\} = \{\}\Delta\{1, 2, 3\} = \{1, 2, 3\},$$

we know the product of all eight elements is simply $(\{1, 2, 3\})^4$, which is equal to $\{\}$,
the identity of the group. For the smaller case of $\Delta(\{1, 2\})$, and all the other groups
also isomorphic to V, the product is the identity as well.

How often does knowing the product of a full row or column help us? If there is
a collection of cages shaped mostly like a row or column but missing a cell, then we
can use the information to determine that cell. In fact, if a collection of cages covers
all cells of any multiple of rows or multiple columns except for a cell, we can find
that cell as well. It helps to search the board for such opportunities at the start of a
puzzle.

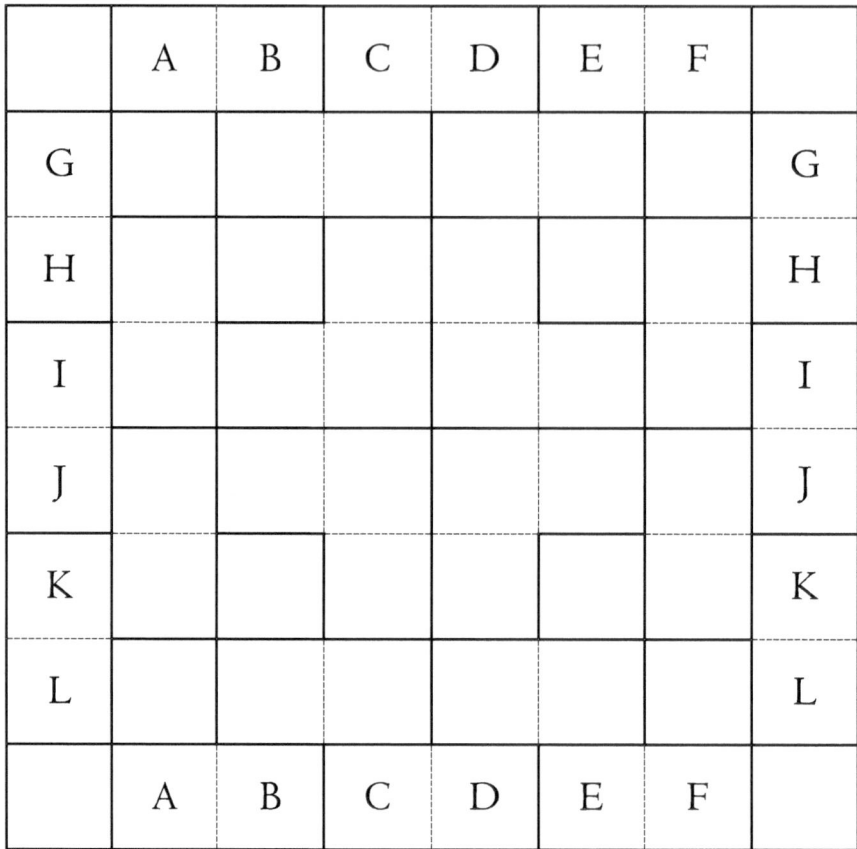

Fig. 5.5 Cells with the same letter are linked so that, knowledge of one immediately gives knowledge of the other

Sometimes the cages cover all the entries in a collection of rows except for two cells. Here, we do not immediately know the values of those cells, but as soon as we find one, we immediately can get the other. Those cells are in some sense linked, and we can ask which compatible pairs in those cells make sense with the rest of the board. For an example, each lettered cell in Fig. 5.5 can be determined from the other cell with the same letter.

For instance, if either cell labeled I is discovered, the other can be determined using the product of the entire top half of the board. If a cell labeled H is discovered, we can determine the value of G in the same cage, and then use the product of the first two rows to find the opposite G and H. Quite often we might not know either of the two cells exactly, but have only a few possibilities for each. In that case, just seeing which pairs of possibilities are compatible with this technique is enough to limit both cells to only one choice.

5.2   Squares and the Number of Possibilities for Adjacent Cells

Once we've exhausted the techniques of the last section, where else in a puzzle might we proceed? Are some cages more productive than others based on the clues alone? We start by considering the most simple possibility: a cage with just two cells and a clue equal to some element c. Let's try to count the possibilities for the elements that we can place in those two cells.

Suppose we consider placing an element a in the leftmost or topmost cell, depending on the orientation of the cage. The other cell must contain the solution to the equation $ax = c$. In a group we can simply multiply both sides on the left by a^{-1} to see that the only solution is $x = a^{-1}c$. Solutions must also satisfy the Latin square condition so if $a^{-1}c = a$, then a is not a candidate for that cell. Left multiplication by a shows this happens exactly when $c = a^2$, so the number of such a limit the total number of possibilities for a two-cell cage with clue c.

Borrowing a term from normal arithmetic, we define a *root* of the element c in our group to be any solution to the equation $x^2 = c$. Borrowing a term from the subject of number theory, we also refer to any group element that does have roots as a *quadratic residue*. Because there are fewer possibilities to consider, cages involving these residues should be the focus of our attention earlier on. The collection of residues is easy to find with a group table, as these are exactly the elements listed along the main diagonal. If r is the number of roots of c, then a two-cell cage with clue c has exactly $n - r$ possibilities, where n is the total number of elements in the group.

Let's examine how this can help us, starting with $\mathbb{Z}_4$. Here our group product is actually addition, so the quadratic residues are the numbers 0 and 2 since

$$0 + 0 = 2 + 2 = 0$$

and

$$1 + 1 = 3 + 3 = 2.$$

This implies that the number of solutions for a cage with two cells equals $4 - 2 = 2$ if the clue c is even, but equals 4 is c is odd. If we have a clue of $c = 0$, those cells must contain 1 and 3 in some order, and if $c = 2$, then those cells must contain 0 and 2 in some order. If c is 1 or 3, then it is possible for any entry to be in either of the cells, though the entry we pick still always determines the other cell.

We can see this in use in the puzzle shown in Fig. 5.6. There we only have two two-cell cages, but the clues are both residues, and in $\mathbb{Z}_4$, this determines the values of their cells. The one-cell cages then help determine their positions, providing enough information to ensure a unique solution that the reader can complete.

Over the Klein four-group V, the square of every single element is the identity, as we have $a^2 = b^2 = c^2 = e$. Thus we have 4 possibilities for a two cell cage if the clue is not e, but 0 possibilities if the clue is e. This might not seem very useful.

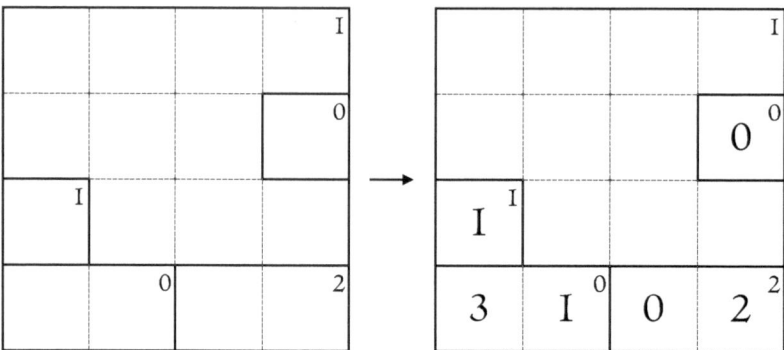

Fig. 5.6 A puzzle over $\mathbb{Z}_4$

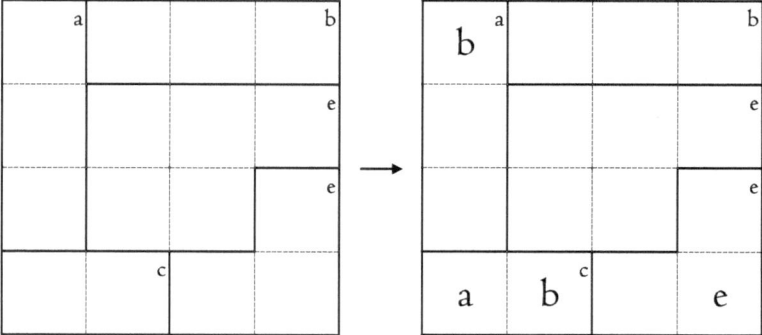

Fig. 5.7 A puzzle over V

After all this book only contains puzzles with solutions, so how does it benefit us to know that a clue of e in a two-cage cell makes for impossible puzzles?

Consider a one-by-three or three-by-one cage with clue x. None of its entries can be x, as that would require the other two entries in the same row or column to have product e. Of course, with this cage shape we could also determine this using the techniques of the last section, but there is still more to learn here. What about an L-shaped cage with clue x? There, the only cell which can contain an x would be the corner, as otherwise the other two entries would be in the same row or column, and again be forced to have product e.

In this Klein four puzzle shown in Fig. 5.7, the three-by-one cages cannot contain their clues in any of their cells, which allows us to place entries in the top left and bottom right corners. The bottom right cage can only contain its clue of e in the corner, which is also useful for completing the puzzle.

In the group $\mathbb{Z}_5$, each number is a quadratic reside with a single root, as $0 = 0 + 0, 2 = 1 + 1, 4 = 2 + 2, 1 = 3 + 3$, and $3 = 4 + 4$. This allows a lot of room for possibilities, with every clue allowing for four possible configurations. For example, a clue of 1 could come from a 0 and 1 in either order, or a 2 and 4 in either order, for a total of four choices. Every clue for a two-cell cage has this same large number of possibilities, and we see that no clue reveals more than any others in this group, at least without considering other parts of the puzzle.

For any odd n, the group $\mathbb{Z}_n$ will have this same behavior. Since we get $0, 2, \cdots, n-1$ from doubling $1, 2, \cdots, \frac{n-1}{2}$, and $1, 3, \cdots, n-2$ from doubling $\frac{n+1}{2}, \frac{n+3}{2}, \cdots, n-1$, we find that in both the even and odd cases, every two cage cell has exactly one prohibited value. For odd c, this value is $\frac{c+n}{2}$ and for even c it is simply $\frac{c}{2}$.

The same is not true for $\mathbb{Z}_n$ when n is even. Here, every odd clue in a two cage allows for a full n possibilities. However, the even clue c forbids both $c/2$ and $(n+c)/2$ since both of these become c when doubled. For example, in $\mathbb{Z}_6$, a two cell cage with a clue of 4 can contain any numbers except for 2 and 5.

This method also leads to some powerful techniques for the non-abelian groups in this book. In any dihedral group, any element of the form $r^k f$ is its own inverse. This means a clue of e in a two cell cage, has very few possibilities. For D_3, the only options are r and r^2 in some order. Since r^2 is its own inverse in D_4, the only possibilities are r and r^3. For the quaternion group, the situation is similar, if not even more restrictive. Every element except 1 and -1 squared is equal to -1. Thus -1 has $8 - 6 = 2$ possibilities, leaving only 1 and -1 in some order to fill the cells of that cage.

Considering the cages with the most restrictive clues first, by examining what the squares are in our group, is a useful early tactic for solving group theory puzzles.

5.3   Tricks with Subgroups and Cosets

We begin this section with an effective strategy for solving puzzles over the group $\mathbb{Z}_4$ which we will later generalize to arbitrary puzzles. The basic idea is that exactly half the cells must be even, and half the cells must be odd. If we can determine where the odd and even cells are, we reduce the choices for each cell down to two possibilities.

Let E be the set $\{0, 2\}$ of evens, and D be the set $\{1, 3\}$ of odds. We choose 'D' for the word 'odd' to avoid any confusion between the letter 'O' and number zero. If we have two elements in E, their sum is also in E. This is a property that no other two element subset of $\{0, 1, 2, 3\}$ has. Also, E turns out to be a group. It inherits associativity from $\mathbb{Z}_4$, contains the identity 0, and both elements are their own inverses. Its tiny table is shown in Fig. 5.8.

Whenever a collection of some of the elements in a group happens to also form a group, we call that collection a *subgroup*. Given any group G, the set containing

Fig. 5.8 A group on the set $\{0, 2\}$

+	0	2
0	0	2
2	2	0

only the identity is a subgroup that we call the *trivial subgroup*. The set containing all of G is also always a subgroup. Any subgroup which does not contain all of G is called a *proper subgroup*. Throughout this section we will focus on non-trivial proper subgroups.

Given any subgroup H of size k and any element a in G, we can form a new set we call $a \cdot H$, by taking the collection of products $a \cdot h$ for every element h in H. This will produce another set of size k, as each of these products will be distinct. This is because ah_1 equals ah_2 exactly when $a^{-1}ah_1 = a^{-1}ah_2$ or $h_1 = h_2$. A set of this type is called a *coset* of the subgroup H, or just a coset if H is understood.

Let's see what the cosets are of the subgroup E in $\mathbb{Z}_4$. We just introduced the notation $a \cdot H$ above for cosets, but since our operation is addition and uses the plus sign, we'll replace our dot with a plus here. As a can be 0, 1, 2 or 3, we have four possibilities: $0 + H$, $1 + H$, $2 + H$, and $3 + H$. If we take either $a = 0$ or $a = 2$, we get a coset that is equal to E since

$$0 + E = \{0 + 0, 0 + 2\} = \{0, 2\} = E$$

and

$$2 + E = \{2 + 0, 2 + 2\} = \{2, 0\} = \{0, 2\} = E.$$

If a equals 1 or 3 the coset turns out to equal D as

$$1 + E = \{1 + 0, 1 + 2\} = \{1, 3\} = D$$

and

$$3 + E = \{3 + 0, 3 + 2\} = \{3, 1\} = \{1, 3\} = E.$$

Thus our subgroup E produced two distinct cosets, E and D. Notice that D is not a subgroup. It does not have an identity, and the "product" of addition in D does not even give us outputs in D, so we cannot even make a table using elements from D alone.

In general, regardless of what our original group G is, and regardless of what our subgroup H is, there will always be exactly $|G|/|H|$ distinct cosets of H, where $|G|$ represents the number of elements in G and $|H|$ represents the number of elements in H. Moreover, every element from G will appear in exactly one of these distinct cosets. The proof of these results is a beautiful part of a first abstract algebra class, but what we want to focus on here is the application of these cosets to solving puzzles.

Let's return to the case of the two cosets for the group $\mathbb{Z}_4$. Not only is the sum of elements in E also in E, but the sum of any two elements in D is also in E. Also the sum of an element in E and an element in D is always in D. The equations

$$E + E = D + D = E$$

Fig. 5.9 A parity group

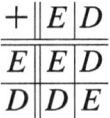

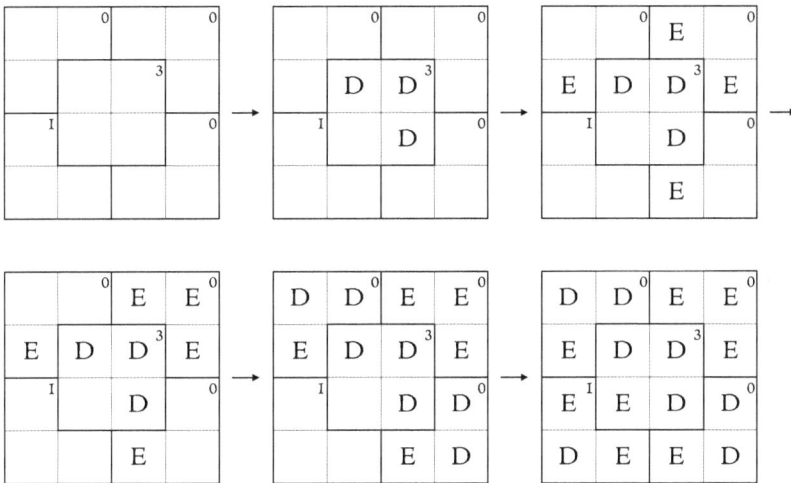

Fig. 5.10 Finding the coset of each cell for a puzzle over $\mathbb{Z}_4$ and thus greatly reducing the number of possibilities

and

$$E + D = D + E = D$$

summarize these rules. It does not matter which element we take from these cosets, the outcome after our group operation will always be in the same coset.

We could even represent these rules in a table that in itself forms a group isomorphic to $\mathbb{Z}_2$, as shown in Fig. 5.9. A table formed from cosets in this manner will always define a group if the original group is abelian. Mathematicians often use the word *parity* to refer to the evenness or oddness of a number, so one might refer to this as a parity group.

Let's now try to apply some of this theory to a specific puzzle, such as the one shown in Fig. 5.10. First, let's examine the top half of the board. Notice that the two cages intersecting the top row have clues of 0. Also, since

$$0 + 1 + 2 + 3 + 0 + 1 + 2 + 3 = 0$$

the entries in the entire top half of the board must sum to zero. For this to happen, we need the two middle entries in the second row to sum to zero as well. The only way two distinct elements of $\mathbb{Z}_4$ can sum to zero is if they are both odd. This same reasoning can be applied to the right half of the board to determine that the two middle entries in the third column must also be odd.

Since each row contains two odds and two evens, we can fill in the rest of the second row and third column. The top right cell is now determined, as that must be even for the clue to be zero. From there, the parity of the rest of the board is quickly determined.

We have actually done more than just reduce each cell to two possibilities. Once we know the entry of any even, we know the entry of the other even in the same row or column. The same is true for the odds. The way the board is configured, any one even or odd entirely determines the rest of the entries of the same parity. There are now only two decisions to be made for the entire puzzle, showing the effectiveness of this technique in certain situations.

Looking in the top left cell, we find it contains two odds in the same row. They must be distinct so they must equal 1 and 3. Since the cage clue is 0 and $1 + 3 = 0$, the even cell in that cage must be 0. This instantly determines the values of all the evens in the puzzle. We can then use the clue in the bottom left cage to find the value of the bottom left odd cell, which then determines the entire board. Fortunately for the integrity of our puzzles, this technique doesn't always lead to an instant answer like it did here. The particular cage pattern in this puzzle made it extra susceptible to this sort of attack, though there is often some information to be gained when looking at cosets [15].

Let's look next at the Klein-four example shown in Fig. 5.11. Unlike $\mathbb{Z}_4$ where there is only one nontrivial proper subgroup, we now have many choices since $\{e, a\}$, $\{e, b\}$, and $\{e, c\}$ all form subgroups. Here we can actually pick any one we like, depending on what we think best suits the clues of our puzzle. For this first example

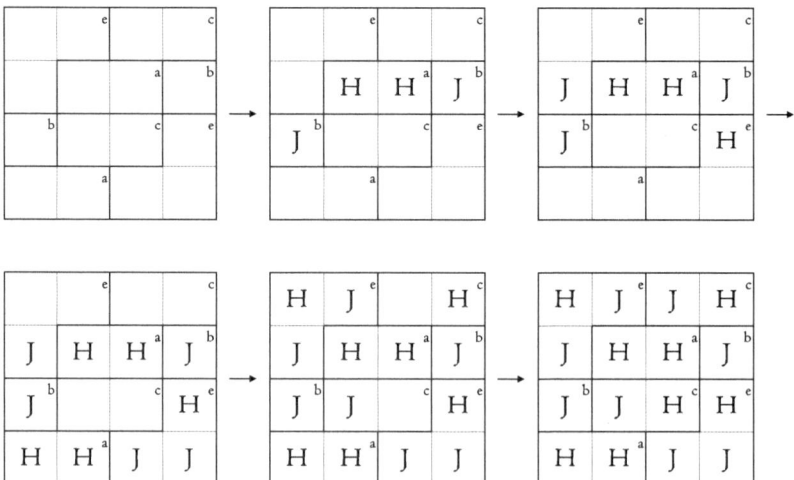

Fig. 5.11 Finding the coset of each cell for a puzzle over V

Fig. 5.12 A two coset group

$$
\begin{array}{c|cc}
\times & H & J \\
\hline
H & H & J \\
J & J & H
\end{array}
$$

let us set $H = \{e, a\}$. The two cosets are then H and the set $J = \{b, c\}$. If we multiply elements of these cosets, then the rules

$$HH = JJ = H$$

and

$$HJ = HJ = J$$

both hold. As we've done before, we can turn these coset rules into a table, producing the result seen in Fig. 5.12.

Look first at the cage in the center of the second row with clue a, which is an element of H. Our rules imply those entries must either be both from H or both from J. Since the last entry in that row is in J, they must both be in H. Now examine row three. The cage in the center must contain one entry from H and one from J. As there is a J in the leftmost cell, this means the last cell in this row must be an H.

In the bottom row, the first two entries must be from the same coset, as the clue is from H. They cannot both be in J, as there are already two J's in the first column, so those cells both contain elements from H. The rest of that row now must contain elements from J. From here, the fact that each row and column has the same number of elements from each coset, is enough to determine the cosets each cell belongs to for the rest of the board.

Once we have matched each cell with the correct coset, we can solve this puzzle quickly. The two entries in the bottom right cage are both from J, so they must be b and c. Together their product is a. The only solution to $ax = e$ is $x = a$ so the entry in row three, column four is a. That allows us to determine the value of every cell from H. The entries from J are given by the two one-cell cages, allowing us to solve the puzzle.

Had we chosen a different subgroup such as $H = \{e, c\}$ we could have completed the board in a very similar fashion. With this different H, it might help to start off examining the cage in the center of row three. We encourage the reader to try it, and the correct coset for each cell is revealed in the solutions section at the end of this book.

If we had chosen $H = \{e, b\}$ to start, then the process of filling out the cosets for each cell would not be quite as straightforward. Part of using this method is learning which cosets will be most effective.

For this group V, regardless of which H we pick, the rules for multiplying elements from H and J are actually the same. The coset multiplication rules, regardless of the H selected, are the same as those in Fig. 5.12.

Fig. 5.13 A three coset
group

$+$	H	I	J
H	H	I	J
I	I	J	H
J	J	H	I

With the group $\mathbb{Z}_6$, there are two nontrivial proper subgroups we could choose for our H. One is the two element group $\{0, 3\}$, and the other is the three element group $\{0, 2, 4\}$. If we set E to be the subgroup $\{0, 2, 4\}$ then we get one other distinct coset, which we can call $D = \{1, 3, 5\}$. If we consider all possible products of elements from E and D, we find our parity rules to be the same as they were for $\mathbb{Z}_4$, and thus we get the table shown in Fig. 5.9 once again. This is also equivalent to the table in Fig. 5.12, as all size two groups are isomorphic. If we set $H = \{0, 3\}$, and label our other two cosets $I = \{1, 4\}$ and $J = \{2, 5\}$, we instead get the table shown in Fig. 5.13, which is isomorphic to the group $\mathbb{Z}_3$.

Which H is best to choose depends greatly on both the clues and cage pattern of the puzzle. We encourage the reader to consider the coset tricks revealed in this section, and to consider how different choices of our subgroup may determine the effectiveness of these techniques. In some situations, we may not be able to figure out which coset every cell belongs to, but we can still get enough information to push ahead in the puzzle, and then use other tricks to take us the rest of the way.

5.4   Using Cosets with Non-abelian Groups

Let's try to use the techniques of the last section for the group D_3. We begin by picking the subgroup $H = \{e, f\}$. We get the three cosets

$$H = eH = fH = \{e, f\},$$

$$I = rH = (rf)H = \{r, rf\},$$

and

$$J = r^2 H = (r^2 f)H = \{r^2, r^2 f\}.$$

We might hope to find rules for multiplying cosets like we did before, but we soon see this is not possible.

What happens if we multiply two elements from I? If we take $r \times r$ we get r^2, which is in J, but if we take $rf \times r$ we get f, which is in H. This shows that the coset we end up in depends on more than just the cosets of the two elements that we are multiplying. Unlike the examples of the last section, here we must consider what the individual elements are. This means we cannot make a rule for which coset $I \times I$ is in, cannot make a coset multiplication table as we did in Figs. 5.9, 5.12, and 5.13, and cannot determine which coset different cells of a puzzle will fall into, like we did in Figs. 5.10 and 5.11.

What went wrong? We have to be a bit more careful with cosets in non-abelian groups. With this particular subgroup H if we multiply the elements on the left by some element a, we often end up in a different coset than when we multiply on the right by a. We defined the coset aH to be the collection of all elements of the form ah where h is in H. In a non-abelian group such as this, the definition differs from the collection of all elements of the form ha where h is in H. Because these are different, let's call sets of the form aH the *left cosets* of H and sets of the form Ha the *right cosets* of H.

Fortunately, we do not need to give up completely on using cosets for D_3 puzzles, though we do need to be more careful. It turns out that with some subgroups, we can multiply cosets, while with others, we cannot. In order to multiply cosets, we will need the left coset aH to give us the same set as the right coset Ha, for any element a in our group. This does not mean that $ax = xa$ for each of the individual elements of H, only that if we multiply each element of H on the left by a, we get the same set as we do from multiplying each element on the right by a.

Let's try again on D_3 using a different subgroup. Take $H = \{e, r, r^2\}$, which gives us $J = \{f, rf, r^2f\}$ as our other coset. Here if we multiply e, r, and r^2 on the left by f, we get the elements f, r^2f and rf in that order. If we multiply on the right by f we get the elements f, rf and r^2f in that order. Even though the individual products might be different, the set they generate in both cases is J. This isn't only because we chose the element f; for any element a, the left and right cosets aH and Ha will be equal. This is the property that we want. Because of it, regardless of which elements in our cosets we are multiplying, the coset of the outcome can always be determined.

To distinguish which subgroups have a useful idea of coset multiplication, we give them a formal name. If H is a subgroup so that $Ha = aH$ for every a in a group G, we call H a *normal subgroup* of G. In the case of abelian groups, every subgroup is normal, so we can use coset tables to help us for any choice of subgroup.

One might now be wondering whether or not it is really worth the time to check which subgroups are normal and which are not. It takes some amount of time just to find all the subgroups, never mind the time spent checking to see which are normal. Fortunately, there are only three different non-abelian groups in this book, so for these cases we can simply reveal all non-trivial proper subgroups here.

For D_3 there are four possible choices: $\{e, f\}$, $\{e, rf\}$, $\{e, r^2f\}$, and $\{e, r, r^2\}$. Only the last of these is normal. For D_4 we have the subgroups $\{e, f\}$, $\{e, rf\}$, $\{e, r^2f\}$, $\{e, r^3f\}$, $\{e, r^2\}$, and $\{e, r, r^2, r^3\}$, the last two of which are normal. For the quaternion group, every subgroup is normal. Non-abelian groups with this property are called *Hamiltonian groups*.[2] The non-trivial proper subgroups are $\{1, -1\}$, $\{1, i, -1, -i\}$, $\{1, j, -1, -j\}$, $\{1, k, -1, -k\}$. Any of these is sometimes a good

---

[2] Since there is only one such group in this book, this not the most import definition to memorize, but we include it because it emphasizes the uniqueness of Q.

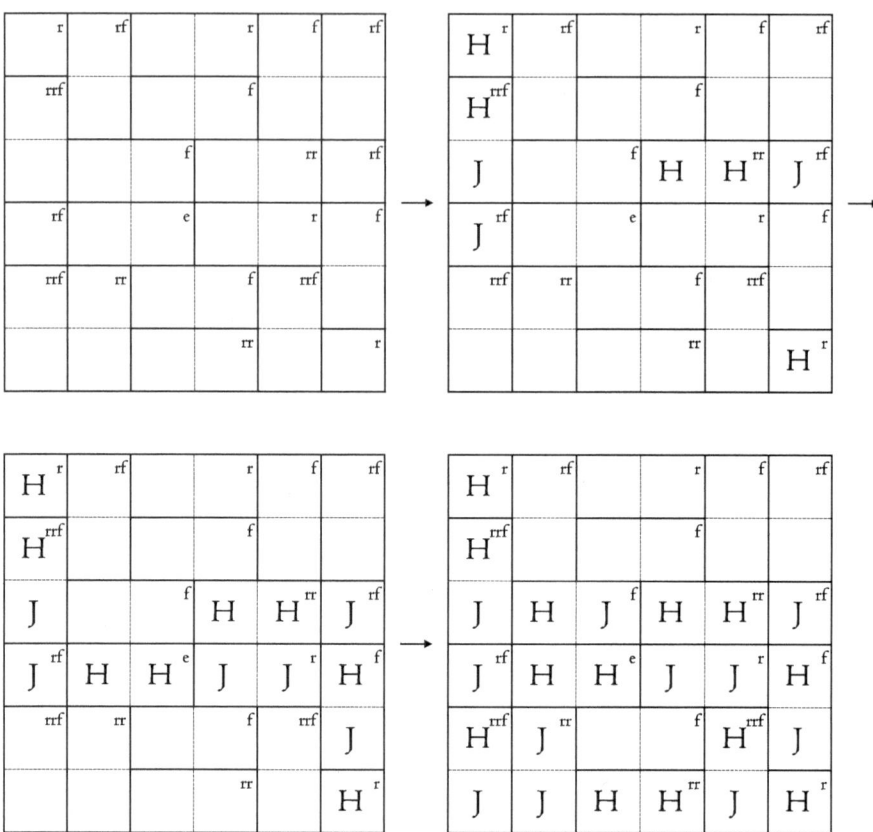

Fig. 5.14 Finding the coset of each cell for a puzzle over D_3

choice for simplifying the puzzle, but as usual, which one we might wish to use comes down to both the cage pattern and clues.

Let's end this last section with an example involving cosets and D_3. Taking $H = \{e, r, r^2\}$ as our subgroup, our other distinct coset becomes $J = \{f, rf, r^2 f\}$. Once again, we get the picture shown in Fig. 5.12 for our coset multiplication table. Let's see how we can use these cosets to help with the puzzle shown in Fig. 5.14.

First, let's examine row three. As f is in J, the two-cell cage with clue f must contain one H and one J, which leaves only one more J for the row. The other cage contained in this row has a clue of rr, which is in H, meaning its entries are both from the same coset. As two J's must already be placed elsewhere, these entries must both be in H. This also determines the coset for the cell in column one, which must be J. That then reveals the coset of the cell above it.

Now turn our attention to column five. The two cells contained in the column have clues in J, so each contains an element from both cosets, leaving one H and one J left to be placed in this column. As we've already placed an H, the coset for row four and column five must be J. The entry to the left must also be in J. Now all J's have been placed in row four, so the remaining cells have entries from H. The entry in row five and column six must also be from J, as the product of that cage is in J.

Starting by examining the cosets of the top cage in column two, we should be able to further fill in our cosets to the final point shown in the figure. At this point, we may not know which coset each element falls into yet, but we've still reduced the number of choices for the majority of cells in our puzzle from six down to three.

In this case, we have three types of cells: ones for which we do not yet know the coset, and others belonging to one of two cosets. It can help to use different colors to shade in cells that we know belong to specific cosets, instead of writing down the coset name. This allows us to keep all our coset information while giving us more room to work inside each cell.

We hope that some of the techniques from this chapter arise naturally from attempting the various puzzles in this book. Perhaps it starts with just a realization that there are lots of cells which can only be even or only be odd, or cells which either must contain a flip or cannot contain a flip. From these observations, the coset techniques presented here are a logical next step. Regardless of how they are reached, these methods are highly effective.

Solutions

6

6.1 Chapter 1 Solutions

Addition

$$\begin{bmatrix} 2 & 1 & 3 & 4 \\ 4 & 3 & 1 & 2 \\ 1 & 4 & 2 & 3 \\ 3 & 2 & 4 & 1 \end{bmatrix} \quad \begin{bmatrix} 3 & 4 & 2 & 1 \\ 2 & 1 & 4 & 3 \\ 1 & 2 & 3 & 4 \\ 4 & 3 & 1 & 2 \end{bmatrix}$$

$$\begin{bmatrix} 1 & 3 & 4 & 2 \\ 2 & 1 & 3 & 4 \\ 4 & 2 & 1 & 3 \\ 3 & 4 & 2 & 1 \end{bmatrix} \quad \begin{bmatrix} 2 & 1 & 3 & 4 \\ 4 & 2 & 1 & 3 \\ 3 & 4 & 2 & 1 \\ 1 & 3 & 4 & 2 \end{bmatrix}$$

$$\begin{bmatrix} 1 & 4 & 2 & 3 \\ 4 & 3 & 1 & 2 \\ 3 & 2 & 4 & 1 \\ 2 & 1 & 3 & 4 \end{bmatrix} \quad \begin{bmatrix} 1 & 3 & 4 & 2 \\ 2 & 4 & 1 & 3 \\ 3 & 1 & 2 & 4 \\ 4 & 2 & 3 & 1 \end{bmatrix}$$

$$\begin{bmatrix} 3 & 1 & 4 & 2 \\ 4 & 3 & 2 & 1 \\ 2 & 4 & 1 & 3 \\ 1 & 2 & 3 & 4 \end{bmatrix} \quad \begin{bmatrix} 2 & 1 & 4 & 3 \\ 3 & 4 & 2 & 1 \\ 1 & 2 & 3 & 4 \\ 4 & 3 & 1 & 2 \end{bmatrix}$$

© The Author(s), under exclusive license to Springer Nature Switzerland AG 2024   137
D. Nacin, *The Group Theory Puzzle Book*, SUMS Readings,
https://doi.org/10.1007/978-3-031-70091-0_6

$$
\begin{bmatrix} 4\,3\,1\,2 \\ 3\,1\,2\,4 \\ 2\,4\,3\,1 \\ 1\,2\,4\,3 \end{bmatrix}
\quad
\begin{bmatrix} 2\,1\,3\,4 \\ 4\,2\,1\,3 \\ 1\,3\,4\,2 \\ 3\,4\,2\,1 \end{bmatrix}
$$

$$
\begin{bmatrix} 4\,2\,3\,1 \\ 3\,4\,1\,2 \\ 1\,3\,2\,4 \\ 2\,1\,4\,3 \end{bmatrix}
\quad
\begin{bmatrix} 3\,2\,4\,1 \\ 1\,3\,2\,4 \\ 4\,1\,3\,2 \\ 2\,4\,1\,3 \end{bmatrix}
$$

$$
\begin{bmatrix} 3\,5\,4\,6\,1\,2 \\ 6\,4\,2\,1\,5\,3 \\ 4\,1\,5\,2\,3\,6 \\ 5\,2\,6\,3\,4\,1 \\ 1\,6\,3\,5\,2\,4 \\ 2\,3\,1\,4\,6\,5 \end{bmatrix}
$$

$$
\begin{bmatrix} 4\,1\,3\,2\,6\,5 \\ 1\,3\,6\,4\,5\,2 \\ 2\,4\,1\,5\,3\,6 \\ 5\,2\,4\,6\,1\,3 \\ 6\,5\,2\,3\,4\,1 \\ 3\,6\,5\,1\,2\,4 \end{bmatrix}
$$

Multiplication

$$
\begin{bmatrix} 4\,3\,1\,2 \\ 2\,1\,3\,4 \\ 3\,2\,4\,1 \\ 1\,4\,2\,3 \end{bmatrix}
\quad
\begin{bmatrix} 4\,2\,1\,3 \\ 3\,1\,4\,2 \\ 2\,4\,3\,1 \\ 1\,3\,2\,4 \end{bmatrix}
$$

$$
\begin{bmatrix} 4\,3\,1\,2 \\ 2\,1\,3\,4 \\ 3\,2\,4\,1 \\ 1\,4\,2\,3 \end{bmatrix}
\quad
\begin{bmatrix} 3\,1\,2\,4 \\ 2\,4\,1\,3 \\ 1\,3\,4\,2 \\ 4\,2\,3\,1 \end{bmatrix}
$$

$$
\begin{bmatrix} 2\,1\,4\,3 \\ 3\,4\,1\,2 \\ 1\,3\,2\,4 \\ 4\,2\,3\,1 \end{bmatrix}
\quad
\begin{bmatrix} 4\,2\,3\,1 \\ 2\,3\,1\,4 \\ 1\,4\,2\,3 \\ 3\,1\,4\,2 \end{bmatrix}
$$

$$\begin{bmatrix} 1\ 2\ 3\ 4 \\ 2\ 1\ 4\ 3 \\ 4\ 3\ 1\ 2 \\ 3\ 4\ 2\ 1 \end{bmatrix} \quad \begin{bmatrix} 3\ 4\ 2\ 1 \\ 4\ 2\ 1\ 3 \\ 2\ 1\ 3\ 4 \\ 1\ 3\ 4\ 2 \end{bmatrix}$$

$$\begin{bmatrix} 1\ 2\ 4\ 3 \\ 3\ 4\ 1\ 2 \\ 2\ 1\ 3\ 4 \\ 4\ 3\ 2\ 1 \end{bmatrix} \quad \begin{bmatrix} 4\ 1\ 3\ 2 \\ 1\ 3\ 2\ 4 \\ 3\ 2\ 4\ 1 \\ 2\ 4\ 1\ 3 \end{bmatrix}$$

$$\begin{bmatrix} 2\ 1\ 4\ 3 \\ 3\ 3\ 2\ 4 \\ 3\ 4\ 1\ 2 \\ 4\ 2\ 3\ 1 \end{bmatrix} \quad \begin{bmatrix} 3\ 1\ 2\ 4 \\ 1\ 2\ 4\ 3 \\ 2\ 4\ 3\ 1 \\ 4\ 3\ 1\ 2 \end{bmatrix}$$

$$\begin{bmatrix} 1\ 6\ 3\ 2\ 5\ 4 \\ 5\ 3\ 6\ 4\ 1\ 2 \\ 3\ 2\ 4\ 5\ 6\ 1 \\ 6\ 4\ 2\ 1\ 3\ 5 \\ 4\ 5\ 1\ 6\ 2\ 3 \\ 2\ 1\ 5\ 3\ 4\ 6 \end{bmatrix}$$

$$\begin{bmatrix} 3\ 2\ 5\ 6\ 4\ 1 \\ 6\ 1\ 3\ 5\ 2\ 4 \\ 2\ 3\ 4\ 1\ 5\ 6 \\ 5\ 4\ 6\ 3\ 1\ 2 \\ 1\ 6\ 2\ 4\ 3\ 5 \\ 4\ 5\ 1\ 2\ 6\ 3 \end{bmatrix}$$

6.2   Chapter 3 Solutions

Cyclic Groups

$$\begin{bmatrix} 2\ 0\ 3\ 1 \\ 0\ 1\ 2\ 3 \\ 1\ 3\ 0\ 2 \\ 3\ 2\ 1\ 0 \end{bmatrix} \quad \begin{bmatrix} 2\ 1\ 0\ 3 \\ 3\ 0\ 1\ 2 \\ 1\ 3\ 2\ 0 \\ 0\ 2\ 3\ 1 \end{bmatrix} \quad \begin{bmatrix} 2\ 3\ 0\ 1 \\ 0\ 2\ 1\ 3 \\ 3\ 1\ 2\ 0 \\ 1\ 0\ 3\ 2 \end{bmatrix}$$

$$
\begin{bmatrix} 1 & 3 & 0 & 2 \\ 3 & 0 & 2 & 1 \\ 0 & 2 & 1 & 3 \\ 2 & 1 & 3 & 0 \end{bmatrix}
\quad
\begin{bmatrix} 0 & 3 & 1 & 2 \\ 3 & 2 & 0 & 1 \\ 2 & 1 & 3 & 0 \\ 1 & 0 & 2 & 3 \end{bmatrix}
\quad
\begin{bmatrix} 0 & 1 & 2 & 3 \\ 3 & 0 & 1 & 2 \\ 2 & 3 & 0 & 1 \\ 1 & 2 & 3 & 0 \end{bmatrix}
$$

$$
\begin{bmatrix} 2 & 0 & 1 & 3 \\ 0 & 2 & 3 & 1 \\ 1 & 3 & 0 & 2 \\ 3 & 1 & 2 & 0 \end{bmatrix}
\quad
\begin{bmatrix} 1 & 3 & 0 & 2 \\ 0 & 2 & 1 & 3 \\ 3 & 0 & 2 & 1 \\ 2 & 1 & 3 & 0 \end{bmatrix}
\quad
\begin{bmatrix} 0 & 3 & 1 & 2 \\ 1 & 2 & 0 & 3 \\ 3 & 1 & 2 & 0 \\ 2 & 0 & 3 & 1 \end{bmatrix}
$$

$$
\begin{bmatrix} 0 & 1 & 2 & 3 \\ 2 & 3 & 0 & 1 \\ 1 & 2 & 3 & 0 \\ 3 & 0 & 1 & 2 \end{bmatrix}
\quad
\begin{bmatrix} 3 & 1 & 0 & 2 \\ 1 & 0 & 2 & 3 \\ 0 & 2 & 3 & 1 \\ 2 & 3 & 1 & 0 \end{bmatrix}
\quad
\begin{bmatrix} 2 & 3 & 1 & 0 \\ 0 & 1 & 2 & 3 \\ 3 & 2 & 0 & 1 \\ 1 & 0 & 3 & 2 \end{bmatrix}
$$

$$
\begin{bmatrix} 3 & 2 & 1 & 0 \\ 0 & 1 & 3 & 2 \\ 1 & 0 & 2 & 3 \\ 2 & 3 & 0 & 1 \end{bmatrix}
\quad
\begin{bmatrix} 2 & 0 & 1 & 3 \\ 0 & 3 & 2 & 1 \\ 3 & 1 & 0 & 2 \\ 1 & 2 & 3 & 0 \end{bmatrix}
\quad
\begin{bmatrix} 1 & 2 & 3 & 0 \\ 3 & 0 & 2 & 1 \\ 0 & 3 & 1 & 2 \\ 2 & 1 & 0 & 3 \end{bmatrix}
$$

$$
\begin{bmatrix} 2 & 0 & 1 & 3 \\ 3 & 1 & 0 & 2 \\ 0 & 3 & 2 & 1 \\ 1 & 2 & 3 & 0 \end{bmatrix}
\quad
\begin{bmatrix} 2 & 0 & 3 & 1 \\ 0 & 3 & 1 & 2 \\ 3 & 1 & 2 & 0 \\ 1 & 2 & 0 & 3 \end{bmatrix}
\quad
\begin{bmatrix} 0 & 2 & 1 & 3 \\ 2 & 1 & 3 & 0 \\ 1 & 3 & 0 & 2 \\ 3 & 0 & 2 & 1 \end{bmatrix}
$$

$$
\begin{bmatrix} 1 & 0 & 3 & 2 \\ 3 & 2 & 0 & 1 \\ 0 & 1 & 2 & 3 \\ 2 & 3 & 1 & 0 \end{bmatrix}
\quad
\begin{bmatrix} 2 & 1 & 3 & 0 \\ 3 & 0 & 1 & 2 \\ 1 & 2 & 0 & 3 \\ 0 & 3 & 2 & 1 \end{bmatrix}
\quad
\begin{bmatrix} 2 & 3 & 0 & 1 \\ 3 & 0 & 1 & 2 \\ 0 & 1 & 2 & 3 \\ 1 & 2 & 3 & 0 \end{bmatrix}
$$

$$
\begin{bmatrix} 0 & 2 & 1 & 3 \\ 3 & 0 & 2 & 1 \\ 1 & 3 & 0 & 2 \\ 2 & 1 & 3 & 0 \end{bmatrix}
\quad
\begin{bmatrix} 3 & 1 & 2 & 0 \\ 0 & 3 & 1 & 2 \\ 2 & 0 & 3 & 1 \\ 1 & 2 & 0 & 3 \end{bmatrix}
\quad
\begin{bmatrix} 3 & 1 & 0 & 2 \\ 1 & 0 & 2 & 3 \\ 0 & 2 & 3 & 1 \\ 2 & 3 & 1 & 0 \end{bmatrix}
$$

$$\mathbb{Z}_5$$

$$\begin{bmatrix} 1\ 2\ 3\ 4\ 0 \\ 2\ 3\ 4\ 0\ 1 \\ 3\ 4\ 0\ 1\ 2 \\ 4\ 0\ 1\ 2\ 3 \\ 0\ 1\ 2\ 3\ 4 \end{bmatrix} \quad \begin{bmatrix} 0\ 4\ 2\ 1\ 3 \\ 2\ 3\ 1\ 4\ 0 \\ 1\ 0\ 4\ 3\ 2 \\ 4\ 2\ 3\ 0\ 1 \\ 3\ 1\ 0\ 2\ 4 \end{bmatrix}$$

$$\begin{bmatrix} 2\ 4\ 3\ 1\ 0 \\ 4\ 1\ 2\ 0\ 3 \\ 1\ 0\ 4\ 3\ 2 \\ 0\ 3\ 1\ 2\ 4 \\ 3\ 2\ 0\ 4\ 1 \end{bmatrix} \quad \begin{bmatrix} 4\ 0\ 1\ 3\ 2 \\ 0\ 2\ 3\ 4\ 1 \\ 3\ 4\ 2\ 1\ 0 \\ 2\ 1\ 4\ 0\ 3 \\ 1\ 3\ 0\ 2\ 4 \end{bmatrix}$$

$$\begin{bmatrix} 1\ 3\ 0\ 2\ 4 \\ 0\ 2\ 4\ 1\ 3 \\ 4\ 1\ 3\ 0\ 2 \\ 2\ 4\ 1\ 3\ 0 \\ 3\ 0\ 2\ 4\ 1 \end{bmatrix} \quad \begin{bmatrix} 0\ 4\ 2\ 3\ 1 \\ 3\ 1\ 4\ 2\ 0 \\ 2\ 0\ 1\ 4\ 2 \\ 1\ 2\ 3\ 0\ 4 \\ 4\ 3\ 0\ 1\ 2 \end{bmatrix}$$

$$\begin{bmatrix} 4\ 1\ 0\ 2\ 3 \\ 3\ 0\ 4\ 1\ 2 \\ 2\ 4\ 3\ 0\ 1 \\ 0\ 2\ 1\ 3\ 4 \\ 1\ 3\ 2\ 4\ 0 \end{bmatrix} \quad \begin{bmatrix} 3\ 4\ 2\ 0\ 1 \\ 4\ 0\ 3\ 1\ 2 \\ 0\ 1\ 4\ 2\ 3 \\ 1\ 2\ 0\ 3\ 4 \\ 2\ 3\ 1\ 4\ 0 \end{bmatrix}$$

$$\begin{bmatrix} 4\ 0\ 3\ 2\ 1 \\ 0\ 3\ 1\ 4\ 2 \\ 1\ 2\ 4\ 3\ 0 \\ 3\ 1\ 2\ 0\ 4 \\ 2\ 4\ 0\ 1\ 3 \end{bmatrix} \quad \begin{bmatrix} 3\ 0\ 4\ 1\ 2 \\ 2\ 4\ 3\ 0\ 1 \\ 0\ 2\ 1\ 3\ 4 \\ 1\ 3\ 2\ 4\ 0 \\ 4\ 1\ 0\ 2\ 3 \end{bmatrix}$$

$$\begin{bmatrix} 0\ 2\ 4\ 3\ 1 \\ 1\ 4\ 3\ 0\ 2 \\ 3\ 1\ 2\ 4\ 0 \\ 4\ 0\ 1\ 2\ 3 \\ 2\ 3\ 0\ 1\ 4 \end{bmatrix} \quad \begin{bmatrix} 4\ 1\ 0\ 2\ 3 \\ 2\ 4\ 3\ 0\ 1 \\ 3\ 0\ 4\ 1\ 2 \\ 1\ 3\ 2\ 4\ 0 \\ 0\ 2\ 1\ 3\ 4 \end{bmatrix}$$

$$\begin{bmatrix} 4 & 2 & 0 & 1 & 3 \\ 1 & 4 & 3 & 0 & 2 \\ 2 & 3 & 1 & 4 & 0 \\ 0 & 1 & 2 & 3 & 4 \\ 3 & 0 & 4 & 2 & 1 \end{bmatrix} \quad \begin{bmatrix} 0 & 1 & 3 & 2 & 4 \\ 3 & 0 & 4 & 1 & 2 \\ 1 & 2 & 0 & 4 & 3 \\ 4 & 3 & 2 & 0 & 1 \\ 2 & 4 & 1 & 3 & 0 \end{bmatrix}$$

$$\begin{bmatrix} 2 & 0 & 4 & 1 & 3 \\ 4 & 2 & 1 & 3 & 0 \\ 1 & 4 & 3 & 0 & 2 \\ 0 & 3 & 2 & 4 & 1 \\ 3 & 1 & 0 & 2 & 4 \end{bmatrix} \quad \begin{bmatrix} 0 & 3 & 1 & 4 & 2 \\ 1 & 2 & 4 & 3 & 0 \\ 4 & 0 & 3 & 2 & 1 \\ 3 & 1 & 2 & 0 & 4 \\ 2 & 4 & 0 & 1 & 3 \end{bmatrix}$$

$$\begin{bmatrix} 1 & 3 & 4 & 0 & 2 \\ 3 & 4 & 0 & 2 & 1 \\ 2 & 1 & 3 & 4 & 0 \\ 4 & 0 & 2 & 1 & 3 \\ 0 & 2 & 1 & 3 & 4 \end{bmatrix} \quad \begin{bmatrix} 1 & 3 & 2 & 0 & 4 \\ 2 & 0 & 4 & 1 & 3 \\ 4 & 1 & 3 & 2 & 0 \\ 0 & 4 & 1 & 3 & 2 \\ 3 & 2 & 0 & 4 & 1 \end{bmatrix}$$

$$\begin{bmatrix} 1 & 4 & 3 & 0 & 2 \\ 2 & 0 & 4 & 1 & 3 \\ 0 & 3 & 2 & 4 & 1 \\ 3 & 1 & 0 & 2 & 4 \\ 4 & 2 & 1 & 3 & 0 \end{bmatrix} \quad \begin{bmatrix} 0 & 1 & 4 & 2 & 3 \\ 1 & 3 & 0 & 4 & 2 \\ 2 & 4 & 3 & 1 & 0 \\ 3 & 2 & 1 & 0 & 4 \\ 4 & 0 & 2 & 3 & 1 \end{bmatrix}$$

$$\begin{bmatrix} 0 & 2 & 1 & 4 & 3 \\ 2 & 1 & 3 & 0 & 4 \\ 1 & 3 & 4 & 2 & 0 \\ 4 & 0 & 2 & 3 & 1 \\ 3 & 4 & 0 & 1 & 2 \end{bmatrix} \quad \begin{bmatrix} 2 & 4 & 1 & 3 & 0 \\ 4 & 0 & 2 & 1 & 3 \\ 3 & 1 & 0 & 4 & 2 \\ 0 & 3 & 4 & 2 & 1 \\ 1 & 2 & 3 & 0 & 4 \end{bmatrix}$$

$$\begin{bmatrix} 1 & 3 & 0 & 2 & 4 \\ 0 & 2 & 4 & 1 & 3 \\ 4 & 1 & 3 & 0 & 2 \\ 2 & 4 & 1 & 3 & 0 \\ 3 & 0 & 2 & 4 & 1 \end{bmatrix} \quad \begin{bmatrix} 2 & 0 & 3 & 1 & 4 \\ 4 & 2 & 0 & 3 & 1 \\ 1 & 4 & 2 & 0 & 3 \\ 3 & 1 & 4 & 2 & 0 \\ 0 & 3 & 1 & 4 & 2 \end{bmatrix}$$

$$
\begin{bmatrix} 2&0&1&4&3 \\ 0&3&4&2&1 \\ 4&2&3&1&0 \\ 3&1&2&0&4 \\ 1&4&0&3&2 \end{bmatrix}
\quad
\begin{bmatrix} 1&0&2&3&4 \\ 2&4&0&1&3 \\ 3&2&1&4&0 \\ 0&3&4&2&1 \\ 4&1&3&0&2 \end{bmatrix}
$$

$$
\begin{bmatrix} 4&3&2&1&0 \\ 0&4&1&3&2 \\ 3&1&0&2&4 \\ 1&2&4&0&3 \\ 2&0&3&4&1 \end{bmatrix}
\quad
\begin{bmatrix} 2&4&1&3&0 \\ 3&0&2&4&1 \\ 0&2&4&1&3 \\ 4&1&3&0&2 \\ 1&3&0&2&4 \end{bmatrix}
$$

$$
\begin{bmatrix} 4&2&1&3&0 \\ 3&1&4&0&2 \\ 1&0&2&4&3 \\ 0&4&3&2&1 \\ 2&3&0&1&4 \end{bmatrix}
\quad
\begin{bmatrix} 2&0&4&1&3 \\ 0&3&2&4&1 \\ 1&4&3&0&2 \\ 3&1&0&2&4 \\ 4&2&1&3&0 \end{bmatrix}
$$

$$\mathbb{Z}_6$$

$$
\begin{bmatrix} 0&3&4&5&1&2 \\ 2&5&3&4&0&1 \\ 3&0&1&2&4&5 \\ 1&4&5&3&2&0 \\ 5&2&0&1&3&4 \\ 4&1&2&0&5&3 \end{bmatrix}
\quad
\begin{bmatrix} 3&5&2&4&0&1 \\ 4&3&1&5&2&0 \\ 1&2&4&0&3&5 \\ 2&0&3&1&5&4 \\ 5&4&0&3&1&2 \\ 0&1&5&2&4&3 \end{bmatrix}
$$

$$
\begin{bmatrix} 4&3&5&2&0&1 \\ 1&5&0&4&3&2 \\ 5&1&2&3&4&0 \\ 0&2&4&5&1&3 \\ 3&4&1&0&2&5 \\ 2&0&3&1&5&4 \end{bmatrix}
\quad
\begin{bmatrix} 1&5&3&4&0&2 \\ 4&2&0&3&5&1 \\ 3&1&5&0&2&4 \\ 5&3&1&2&4&0 \\ 2&0&4&1&3&5 \\ 0&4&2&5&1&3 \end{bmatrix}
$$

$$
\begin{bmatrix} 0&5&3&1&4&2 \\ 5&1&2&0&3&4 \\ 2&4&1&3&5&0 \\ 1&0&4&5&2&3 \\ 4&3&0&2&1&5 \\ 3&2&5&4&0&1 \end{bmatrix}
\quad
\begin{bmatrix} 5&2&1&3&4&0 \\ 4&0&3&5&2&1 \\ 2&1&5&4&0&3 \\ 0&3&4&2&1&5 \\ 1&5&2&0&3&4 \\ 3&4&0&1&5&2 \end{bmatrix}
$$

$$
\begin{bmatrix}
5 & 4 & 3 & 1 & 2 & 0 \\
1 & 0 & 5 & 3 & 4 & 2 \\
4 & 5 & 3 & 0 & 3 & 1 \\
2 & 3 & 0 & 4 & 1 & 5 \\
0 & 1 & 4 & 2 & 5 & 3 \\
3 & 2 & 1 & 5 & 0 & 4
\end{bmatrix}
\quad
\begin{bmatrix}
3 & 4 & 2 & 5 & 1 & 0 \\
5 & 0 & 3 & 2 & 4 & 1 \\
2 & 1 & 5 & 3 & 0 & 4 \\
4 & 2 & 1 & 0 & 5 & 3 \\
0 & 3 & 4 & 1 & 2 & 5 \\
1 & 5 & 0 & 4 & 3 & 2
\end{bmatrix}
$$

$$
\begin{bmatrix}
2 & 5 & 1 & 4 & 3 & 0 \\
3 & 0 & 5 & 1 & 4 & 2 \\
4 & 2 & 0 & 5 & 1 & 3 \\
0 & 1 & 4 & 3 & 2 & 5 \\
1 & 3 & 2 & 0 & 5 & 4 \\
5 & 4 & 3 & 2 & 0 & 1
\end{bmatrix}
\quad
\begin{bmatrix}
1 & 4 & 2 & 3 & 5 & 0 \\
2 & 1 & 3 & 5 & 0 & 4 \\
0 & 5 & 4 & 1 & 2 & 3 \\
3 & 2 & 5 & 0 & 4 & 1 \\
5 & 3 & 0 & 4 & 1 & 2 \\
4 & 0 & 1 & 2 & 3 & 5
\end{bmatrix}
$$

$$
\begin{bmatrix}
0 & 4 & 2 & 5 & 1 & 3 \\
1 & 0 & 3 & 2 & 4 & 5 \\
3 & 2 & 1 & 0 & 5 & 4 \\
2 & 5 & 0 & 4 & 3 & 1 \\
5 & 3 & 4 & 1 & 2 & 0 \\
4 & 1 & 5 & 3 & 0 & 2
\end{bmatrix}
\quad
\begin{bmatrix}
5 & 0 & 4 & 1 & 3 & 2 \\
2 & 5 & 1 & 0 & 4 & 3 \\
4 & 3 & 5 & 2 & 0 & 1 \\
3 & 2 & 0 & 5 & 1 & 4 \\
1 & 4 & 2 & 3 & 5 & 0 \\
0 & 1 & 3 & 4 & 2 & 5
\end{bmatrix}
$$

$$\mathbb{Z}_7$$

$$
\begin{bmatrix}
4 & 5 & 0 & 2 & 3 & 1 & 6 \\
1 & 4 & 5 & 3 & 0 & 6 & 2 \\
2 & 6 & 1 & 5 & 4 & 3 & 0 \\
0 & 3 & 2 & 1 & 6 & 5 & 4 \\
6 & 1 & 4 & 0 & 5 & 2 & 3 \\
3 & 2 & 6 & 4 & 1 & 0 & 5 \\
5 & 0 & 3 & 6 & 2 & 4 & 1
\end{bmatrix}
$$

$$
\begin{bmatrix}
2 & 6 & 1 & 5 & 0 & 3 & 4 \\
1 & 3 & 6 & 4 & 5 & 0 & 2 \\
4 & 1 & 2 & 0 & 3 & 6 & 5 \\
0 & 4 & 5 & 6 & 1 & 2 & 3 \\
3 & 5 & 0 & 1 & 2 & 4 & 6 \\
5 & 2 & 4 & 3 & 6 & 1 & 0 \\
6 & 0 & 3 & 2 & 4 & 5 & 1
\end{bmatrix}
$$

$$
\begin{bmatrix}
5 & 1 & 4 & 6 & 3 & 2 & 0 \\
3 & 6 & 1 & 2 & 4 & 0 & 5 \\
2 & 3 & 5 & 4 & 0 & 1 & 6 \\
0 & 4 & 3 & 1 & 5 & 6 & 2 \\
4 & 2 & 6 & 0 & 1 & 5 & 3 \\
1 & 0 & 2 & 5 & 6 & 3 & 4 \\
6 & 5 & 0 & 3 & 2 & 4 & 1
\end{bmatrix}
$$

$$
\begin{bmatrix}
2 & 0 & 5 & 4 & 6 & 1 & 3 \\
1 & 3 & 0 & 2 & 5 & 6 & 4 \\
0 & 1 & 2 & 5 & 4 & 3 & 6 \\
3 & 6 & 1 & 0 & 2 & 4 & 5 \\
6 & 4 & 3 & 1 & 0 & 5 & 2 \\
4 & 5 & 6 & 3 & 1 & 2 & 0 \\
5 & 2 & 4 & 6 & 3 & 0 & 1
\end{bmatrix}
$$

$$\mathbb{Z}_8$$

$$
\begin{bmatrix}
5 & 1 & 3 & 0 & 6 & 7 & 2 & 4 \\
2 & 0 & 6 & 5 & 1 & 4 & 7 & 3 \\
4 & 2 & 0 & 7 & 5 & 6 & 3 & 1 \\
6 & 4 & 2 & 3 & 7 & 0 & 1 & 5 \\
3 & 7 & 5 & 4 & 2 & 1 & 6 & 0 \\
1 & 3 & 7 & 6 & 4 & 5 & 0 & 2 \\
0 & 6 & 4 & 1 & 3 & 2 & 5 & 7 \\
7 & 5 & 1 & 2 & 0 & 3 & 4 & 6
\end{bmatrix}
$$

$$
\begin{bmatrix}
6 & 0 & 4 & 7 & 1 & 2 & 5 & 3 \\
5 & 3 & 1 & 2 & 4 & 7 & 6 & 0 \\
7 & 6 & 0 & 1 & 3 & 4 & 2 & 5 \\
0 & 4 & 2 & 6 & 7 & 5 & 3 & 1 \\
1 & 7 & 6 & 3 & 5 & 0 & 4 & 2 \\
3 & 1 & 7 & 5 & 2 & 6 & 0 & 4 \\
4 & 2 & 5 & 0 & 6 & 3 & 1 & 7 \\
2 & 5 & 3 & 4 & 0 & 1 & 7 & 6
\end{bmatrix}
$$

$$\begin{bmatrix} 3 1 2 4 7 5 0 6 \\ 4 5 7 1 6 2 3 0 \\ 5 7 0 2 3 6 1 4 \\ 1 2 6 5 0 7 4 3 \\ 6 3 1 0 5 4 7 2 \\ 7 0 4 6 1 3 2 5 \\ 2 6 3 7 4 0 5 1 \\ 0 4 5 3 2 1 6 7 \end{bmatrix}$$

$$\begin{bmatrix} 2 4 0 1 7 3 6 5 \\ 7 6 3 4 1 2 5 0 \\ 0 7 6 2 3 5 1 4 \\ 1 5 2 6 4 7 0 3 \\ 6 3 1 0 5 4 2 7 \\ 4 0 7 5 6 1 3 2 \\ 3 1 5 7 2 0 4 6 \\ 5 2 4 3 0 6 7 1 \end{bmatrix}$$

$$\begin{bmatrix} 3 0 6 5 7 2 1 4 \\ 0 1 4 2 6 3 7 5 \\ 1 7 5 3 4 0 6 2 \\ 2 3 7 4 1 5 0 6 \\ 4 5 0 7 3 6 2 1 \\ 7 6 2 0 5 1 4 3 \\ 6 4 3 1 2 7 5 0 \\ 5 2 1 6 0 4 3 7 \end{bmatrix}$$

$$\begin{bmatrix} 6 2 3 7 4 0 1 5 \\ 4 6 0 3 1 2 5 7 \\ 0 3 5 1 2 7 6 4 \\ 1 4 2 0 5 6 7 3 \\ 2 0 7 5 6 3 4 1 \\ 7 5 4 6 3 1 0 2 \\ 3 7 1 4 0 5 2 6 \\ 5 1 6 2 7 4 3 0 \end{bmatrix}$$

More Cyclic Groups

$$U(5)$$

$$\begin{bmatrix} 2\ 3\ 4\ 1 \\ 1\ 2\ 3\ 4 \\ 4\ 1\ 2\ 3 \\ 3\ 4\ 1\ 2 \end{bmatrix} \quad \begin{bmatrix} 4\ 3\ 2\ 1 \\ 2\ 4\ 1\ 3 \\ 1\ 2\ 3\ 4 \\ 3\ 1\ 4\ 2 \end{bmatrix} \quad \begin{bmatrix} 1\ 3\ 4\ 2 \\ 2\ 4\ 1\ 3 \\ 3\ 1\ 2\ 4 \\ 4\ 2\ 3\ 1 \end{bmatrix}$$

$$\begin{bmatrix} 3\ 2\ 1\ 4 \\ 1\ 4\ 3\ 2 \\ 4\ 3\ 2\ 1 \\ 2\ 1\ 4\ 3 \end{bmatrix} \quad \begin{bmatrix} 2\ 4\ 3\ 1 \\ 1\ 3\ 4\ 2 \\ 3\ 2\ 1\ 4 \\ 4\ 1\ 2\ 3 \end{bmatrix} \quad \begin{bmatrix} 3\ 2\ 1\ 4 \\ 2\ 1\ 4\ 3 \\ 1\ 4\ 3\ 2 \\ 4\ 3\ 2\ 1 \end{bmatrix}$$

$$U(10)$$

$$\begin{bmatrix} 1\ 9\ 7\ 3 \\ 9\ 1\ 3\ 7 \\ 3\ 7\ 1\ 9 \\ 7\ 3\ 9\ 1 \end{bmatrix} \quad \begin{bmatrix} 3\ 1\ 9\ 7 \\ 7\ 9\ 1\ 3 \\ 9\ 3\ 7\ 1 \\ 1\ 7\ 3\ 9 \end{bmatrix} \quad \begin{bmatrix} 1\ 9\ 3\ 7 \\ 9\ 3\ 7\ 1 \\ 7\ 1\ 9\ 3 \\ 3\ 7\ 1\ 9 \end{bmatrix}$$

$$\begin{bmatrix} 7\ 3\ 1\ 9 \\ 3\ 1\ 9\ 7 \\ 1\ 9\ 7\ 3 \\ 9\ 7\ 3\ 1 \end{bmatrix} \quad \begin{bmatrix} 3\ 1\ 7\ 9 \\ 1\ 9\ 3\ 7 \\ 9\ 7\ 1\ 3 \\ 7\ 3\ 9\ 1 \end{bmatrix} \quad \begin{bmatrix} 7\ 3\ 1\ 9 \\ 3\ 1\ 9\ 7 \\ 1\ 9\ 7\ 3 \\ 9\ 7\ 3\ 1 \end{bmatrix}$$

$$U(7)$$

$$\begin{bmatrix} 3\ 4\ 6\ 2\ 5\ 1 \\ 5\ 1\ 3\ 4\ 6\ 2 \\ 2\ 5\ 1\ 3\ 4\ 6 \\ 1\ 3\ 4\ 6\ 2\ 5 \\ 4\ 6\ 2\ 5\ 1\ 3 \\ 6\ 2\ 5\ 1\ 3\ 4 \end{bmatrix} \quad \begin{bmatrix} 4\ 6\ 3\ 1\ 2\ 5 \\ 3\ 5\ 1\ 6\ 4\ 2 \\ 2\ 1\ 4\ 3\ 5\ 6 \\ 6\ 4\ 5\ 2\ 1\ 3 \\ 1\ 2\ 6\ 5\ 3\ 4 \\ 5\ 3\ 2\ 4\ 6\ 1 \end{bmatrix}$$

$$U(9)$$

$$\begin{bmatrix} 4\ 7\ 2\ 1\ 5\ 8 \\ 5\ 8\ 1\ 2\ 4\ 7 \\ 1\ 4\ 7\ 8\ 2\ 5 \\ 8\ 1\ 4\ 5\ 7\ 2 \\ 2\ 5\ 8\ 7\ 1\ 4 \\ 7\ 2\ 5\ 4\ 8\ 1 \end{bmatrix} \quad \begin{bmatrix} 5\ 7\ 1\ 4\ 2\ 8 \\ 7\ 4\ 5\ 8\ 1\ 2 \\ 1\ 5\ 2\ 7\ 8\ 4 \\ 8\ 2\ 4\ 1\ 7\ 5 \\ 4\ 8\ 7\ 2\ 5\ 1 \\ 2\ 1\ 8\ 5\ 4\ 7 \end{bmatrix}$$

Other Abelian Groups

U8

$$\begin{bmatrix} 7 & 1 & 3 & 5 \\ 5 & 3 & 7 & 1 \\ 1 & 7 & 5 & 3 \\ 3 & 5 & 1 & 7 \end{bmatrix} \quad \begin{bmatrix} 3 & 1 & 5 & 7 \\ 7 & 3 & 1 & 5 \\ 5 & 7 & 3 & 1 \\ 1 & 5 & 7 & 3 \end{bmatrix} \quad \begin{bmatrix} 7 & 3 & 5 & 1 \\ 1 & 5 & 7 & 3 \\ 3 & 7 & 1 & 5 \\ 5 & 1 & 3 & 7 \end{bmatrix}$$

$$\begin{bmatrix} 5 & 3 & 7 & 1 \\ 7 & 1 & 5 & 3 \\ 1 & 5 & 3 & 7 \\ 3 & 7 & 1 & 5 \end{bmatrix} \quad \begin{bmatrix} 1 & 7 & 3 & 5 \\ 3 & 5 & 1 & 7 \\ 7 & 3 & 5 & 1 \\ 5 & 1 & 7 & 3 \end{bmatrix} \quad \begin{bmatrix} 7 & 1 & 3 & 5 \\ 1 & 3 & 5 & 7 \\ 3 & 5 & 7 & 1 \\ 5 & 7 & 1 & 3 \end{bmatrix}$$

U12

$$\begin{bmatrix} 7 & 1 & 11 & 5 \\ 1 & 5 & 7 & 11 \\ 5 & 11 & 1 & 7 \\ 11 & 7 & 5 & 1 \end{bmatrix} \quad \begin{bmatrix} 7 & 5 & 1 & 11 \\ 11 & 1 & 5 & 7 \\ 1 & 7 & 11 & 5 \\ 5 & 11 & 7 & 1 \end{bmatrix} \quad \begin{bmatrix} 1 & 11 & 7 & 5 \\ 11 & 7 & 5 & 1 \\ 7 & 5 & 1 & 11 \\ 5 & 1 & 11 & 7 \end{bmatrix}$$

$$\begin{bmatrix} 11 & 7 & 1 & 5 \\ 7 & 1 & 5 & 11 \\ 1 & 5 & 11 & 7 \\ 5 & 11 & 7 & 1 \end{bmatrix} \quad \begin{bmatrix} 5 & 7 & 1 & 11 \\ 11 & 5 & 7 & 1 \\ 1 & 11 & 5 & 7 \\ 7 & 1 & 11 & 5 \end{bmatrix} \quad \begin{bmatrix} 5 & 1 & 11 & 7 \\ 11 & 7 & 1 & 5 \\ 1 & 5 & 7 & 11 \\ 7 & 11 & 5 & 1 \end{bmatrix}$$

V

$$\begin{bmatrix} b & c & a & e \\ c & a & e & b \\ a & e & b & c \\ e & b & c & a \end{bmatrix} \quad \begin{bmatrix} c & a & e & b \\ e & c & b & a \\ b & e & a & c \\ a & b & c & e \end{bmatrix} \quad \begin{bmatrix} a & b & c & e \\ e & a & b & c \\ b & c & e & a \\ c & e & a & b \end{bmatrix}$$

$$\begin{bmatrix} e & a & b & c \\ c & b & a & e \\ a & c & e & b \\ b & e & c & a \end{bmatrix} \quad \begin{bmatrix} c & e & a & b \\ b & c & e & a \\ a & b & c & e \\ e & a & b & c \end{bmatrix} \quad \begin{bmatrix} c & a & e & b \\ b & e & a & c \\ a & b & c & e \\ e & c & b & a \end{bmatrix}$$

$$\begin{bmatrix} c & a & b & e \\ b & e & a & c \\ a & c & e & b \\ e & b & c & a \end{bmatrix} \quad \begin{bmatrix} a & c & e & b \\ e & b & a & c \\ b & a & c & e \\ c & e & b & a \end{bmatrix} \quad \begin{bmatrix} a & e & c & b \\ e & a & b & c \\ c & b & e & a \\ b & c & a & e \end{bmatrix}$$

$$\begin{bmatrix} a & b & e & c \\ c & e & b & a \\ b & c & a & e \\ e & a & c & b \end{bmatrix} \quad \begin{bmatrix} a & b & c & e \\ b & c & e & a \\ e & a & b & c \\ c & e & a & b \end{bmatrix} \quad \begin{bmatrix} a & b & c & e \\ e & a & b & c \\ b & c & e & a \\ c & e & a & b \end{bmatrix}$$

$$\begin{bmatrix} a & c & b & e \\ b & e & c & a \\ c & a & e & b \\ e & b & a & c \end{bmatrix} \quad \begin{bmatrix} b & a & e & c \\ a & b & c & e \\ e & c & a & b \\ c & e & b & a \end{bmatrix} \quad \begin{bmatrix} e & c & a & b \\ b & a & e & c \\ c & e & b & a \\ a & b & c & e \end{bmatrix}$$

$$\begin{bmatrix} a & e & b & c \\ e & b & c & a \\ b & c & a & e \\ c & a & e & b \end{bmatrix} \quad \begin{bmatrix} a & c & e & b \\ e & b & a & c \\ c & e & b & a \\ b & a & c & e \end{bmatrix} \quad \begin{bmatrix} e & a & c & b \\ c & b & e & a \\ b & e & a & c \\ a & c & b & e \end{bmatrix}$$

$$\begin{bmatrix} c & a & b & e \\ e & c & a & b \\ b & e & c & a \\ a & b & e & c \end{bmatrix} \quad \begin{bmatrix} e & b & c & a \\ b & c & a & e \\ c & a & e & b \\ a & e & b & c \end{bmatrix} \quad \begin{bmatrix} a & e & b & c \\ c & a & e & b \\ b & c & a & e \\ e & b & c & a \end{bmatrix}$$

$$\begin{bmatrix} e & c & b & a \\ a & e & c & b \\ b & a & e & c \\ c & b & a & e \end{bmatrix} \quad \begin{bmatrix} a & e & b & c \\ e & b & c & a \\ b & c & a & e \\ c & a & e & b \end{bmatrix} \quad \begin{bmatrix} c & a & b & e \\ e & c & a & b \\ b & e & c & a \\ a & b & e & c \end{bmatrix}$$

$$\Delta(\{1, 2\})$$

$$\begin{bmatrix} \{1\} & \{\} & \{1,2\} & \{2\} \\ \{\} & \{1,2\} & \{2\} & \{1\} \\ \{1,2\} & \{2\} & \{1\} & \{\} \\ \{2\} & \{1\} & \{\} & \{1,2\} \end{bmatrix} \begin{bmatrix} \{1\} & \{2\} & \{1,2\} & \{\} \\ \{1,2\} & \{1\} & \{\} & \{2\} \\ \{\} & \{1,2\} & \{2\} & \{1\} \\ \{2\} & \{\} & \{1\} & \{1,2\} \end{bmatrix} \begin{bmatrix} \{1,2\} & \{1\} & \{2\} & \{\} \\ \{2\} & \{\} & \{1,2\} & \{1\} \\ \{\} & \{1,2\} & \{1\} & \{2\} \\ \{1\} & \{2\} & \{\} & \{1,2\} \end{bmatrix}$$

$$\begin{bmatrix} \{1\} & \{2\} & \{\} & \{1,2\} \\ \{\} & \{1,2\} & \{1\} & \{2\} \\ \{2\} & \{\} & \{1,2\} & \{1\} \\ \{1,2\} & \{1\} & \{2\} & \{\} \end{bmatrix} \begin{bmatrix} \{2\} & \{\} & \{1\} & \{1,2\} \\ \{1\} & \{1,2\} & \{2\} & \{\} \\ \{\} & \{1\} & \{1,2\} & \{2\} \\ \{1,2\} & \{2\} & \{\} & \{1\} \end{bmatrix} \begin{bmatrix} \{1\} & \{2\} & \{1,2\} & \{\} \\ \{\} & \{1,2\} & \{1\} & \{2\} \\ \{2\} & \{1\} & \{\} & \{1,2\} \\ \{1,2\} & \{\} & \{2\} & \{1\} \end{bmatrix}$$

$$\mathbb{Z}_2 \times \mathbb{Z}_2$$

$$\begin{bmatrix} (1,0) & (0,0) & (0,1) & (1,1) \\ (0,0) & (0,1) & (1,1) & (1,0) \\ (1,1) & (1,0) & (0,0) & (0,1) \\ (0,1) & (1,1) & (1,0) & (0,0) \end{bmatrix} \begin{bmatrix} (1,0) & (1,1) & (0,0) & (0,1) \\ (1,1) & (0,0) & (0,1) & (1,0) \\ (0,0) & (0,1) & (1,0) & (1,1) \\ (0,1) & (1,0) & (1,1) & (0,0) \end{bmatrix} \begin{bmatrix} (0,1) & (1,0) & (0,0) & (1,1) \\ (1,1) & (0,0) & (0,1) & (1,0) \\ (1,0) & (0,1) & (1,1) & (0,0) \\ (0,0) & (1,1) & (1,0) & (0,1) \end{bmatrix}$$

$$\begin{bmatrix} (1,1) & (1,0) & (0,0) & (0,1) \\ (1,0) & (0,0) & (0,1) & (1,1) \\ (0,0) & (0,1) & (1,1) & (1,0) \\ (0,1) & (1,1) & (1,0) & (0,0) \end{bmatrix} \begin{bmatrix} (0,1) & (1,0) & (0,0) & (1,1) \\ (1,0) & (0,0) & (1,1) & (0,1) \\ (0,0) & (1,1) & (0,1) & (1,0) \\ (1,1) & (0,1) & (1,0) & (0,0) \end{bmatrix} \begin{bmatrix} (0,1) & (1,1) & (0,0) & (1,0) \\ (1,0) & (0,1) & (1,1) & (0,0) \\ (0,0) & (1,0) & (0,1) & (1,1) \\ (1,1) & (0,0) & (1,0) & (0,1) \end{bmatrix}$$

$$\Delta(\{1,2,3\})$$

$$\begin{bmatrix} \{1\} & \{1,2\} & \{2\} & \{2,3\} & \{\} & \{1,3\} & \{3\} & \{1,2,3\} \\ \{2\} & \{1,3\} & \{2,3\} & \{1,2,3\} & \{3\} & \{\} & \{1\} & \{1,2\} \\ \{1,2,3\} & \{3\} & \{1,2\} & \{1,3\} & \{2\} & \{1\} & \{2,3\} & \{\} \\ \{2,3\} & \{\} & \{1,2,3\} & \{1,2\} & \{1\} & \{3\} & \{2\} & \{1,3\} \\ \{1,2\} & \{1\} & \{1,3\} & \{\} & \{2,3\} & \{2\} & \{1,2,3\} & \{3\} \\ \{1,3\} & \{2\} & \{\} & \{3\} & \{1,2,3\} & \{2,3\} & \{1,2\} & \{1\} \\ \{3\} & \{1,2,3\} & \{1\} & \{2\} & \{1,3\} & \{1,2\} & \{\} & \{2,3\} \\ \{\} & \{2,3\} & \{3\} & \{1\} & \{1,2\} & \{1,2,3\} & \{1,3\} & \{2\} \end{bmatrix}$$

$$\begin{bmatrix} \{1,2,3\} & \{2,3\} & \{1,3\} & \{1,2\} & \{2\} & \{1\} & \{\} & \{3\} \\ \{1\} & \{2\} & \{3\} & \{\} & \{2,3\} & \{1,2,3\} & \{1,2\} & \{1,3\} \\ \{1,2\} & \{3\} & \{2,3\} & \{1\} & \{1,3\} & \{\} & \{1,2,3\} & \{2\} \\ \{1,3\} & \{1\} & \{1,2\} & \{2,3\} & \{1,2,3\} & \{3\} & \{2\} & \{\} \\ \{\} & \{1,3\} & \{2\} & \{1,2,3\} & \{3\} & \{1,2\} & \{1\} & \{2,3\} \\ \{2,3\} & \{\} & \{1\} & \{3\} & \{1,2\} & \{2\} & \{1,3\} & \{1,2,3\} \\ \{2\} & \{1,2\} & \{1,2,3\} & \{1,3\} & \{\} & \{2,3\} & \{3\} & \{1\} \\ \{3\} & \{1,2,3\} & \{\} & \{2\} & \{1\} & \{1,3\} & \{2,3\} & \{1,2\} \end{bmatrix}$$

$$\begin{bmatrix} \{3\} & \{1,2,3\} & \{\} & \{1,2\} & \{2\} & \{1,3\} & \{2,3\} & \{1\} \\ \{2,3\} & \{1,3\} & \{1,2\} & \{1\} & \{\} & \{3\} & \{2\} & \{1,2,3\} \\ \{\} & \{2,3\} & \{1,2,3\} & \{1,3\} & \{1\} & \{2\} & \{1,2\} & \{3\} \\ \{1\} & \{\} & \{3\} & \{2,3\} & \{1,3\} & \{1,2\} & \{1,2,3\} & \{2\} \\ \{1,2\} & \{2\} & \{1,3\} & \{3\} & \{1,2,3\} & \{\} & \{1\} & \{2,3\} \\ \{1,2,3\} & \{1,2\} & \{2,3\} & \{2\} & \{3\} & \{1\} & \{1,3\} & \{\} \\ \{2\} & \{3\} & \{1\} & \{1,2,3\} & \{1,2\} & \{2,3\} & \{\} & \{1,3\} \\ \{1,3\} & \{1\} & \{2\} & \{\} & \{2,3\} & \{1,2,3\} & \{3\} & \{1,2\} \end{bmatrix}$$

$$\begin{bmatrix}
\{1,2\} & \{1,2,3\} & \{1\} & \{1,3\} & \{3\} & \{2\} & \{\} & \{2,3\} \\
\{1\} & \{1,2\} & \{2,3\} & \{3\} & \{2\} & \{1,2,3\} & \{1,3\} & \{\} \\
\{1,2,3\} & \{2\} & \{1,2\} & \{\} & \{1,3\} & \{3\} & \{2,3\} & \{1\} \\
\{2\} & \{3\} & \{1,2,3\} & \{2,3\} & \{\} & \{1,3\} & \{1\} & \{1,2\} \\
\{1,3\} & \{\} & \{3\} & \{1,2\} & \{1\} & \{2,3\} & \{1,2,3\} & \{2\} \\
\{2,3\} & \{1\} & \{\} & \{2\} & \{1,2,3\} & \{1,2\} & \{3\} & \{1,3\} \\
\{\} & \{2,3\} & \{1,3\} & \{1,2,3\} & \{1,2\} & \{1\} & \{2\} & \{3\} \\
\{3\} & \{1,3\} & \{2\} & \{1\} & \{2,3\} & \{\} & \{1,2\} & \{1,2,3\}
\end{bmatrix}$$

$$\begin{bmatrix}
\{2,3\} & \{\} & \{3\} & \{1,2\} & \{1,2,3\} & \{2\} & \{1\} & \{1,3\} \\
\{3\} & \{1,3\} & \{1,2\} & \{1,2,3\} & \{\} & \{2,3\} & \{2\} & \{1\} \\
\{\} & \{2,3\} & \{1,3\} & \{1\} & \{2\} & \{1,2,3\} & \{1,2\} & \{3\} \\
\{1\} & \{1,2\} & \{2\} & \{2,3\} & \{3\} & \{1,3\} & \{\} & \{1,2,3\} \\
\{1,2\} & \{1\} & \{1,2,3\} & \{\} & \{1,3\} & \{3\} & \{2,3\} & \{2\} \\
\{1,3\} & \{3\} & \{1\} & \{2\} & \{2,3\} & \{\} & \{1,2,3\} & \{1,2\} \\
\{1,2,3\} & \{2\} & \{\} & \{1,3\} & \{1\} & \{1,2\} & \{3\} & \{2,3\} \\
\{2\} & \{1,2,3\} & \{2,3\} & \{3\} & \{1,2\} & \{1\} & \{1,3\} & \{\}
\end{bmatrix}$$

$$\begin{bmatrix}
\{2,3\} & \{1,2\} & \{2\} & \{\} & \{1,3\} & \{3\} & \{1\} & \{1,2,3\} \\
\{1,2\} & \{2\} & \{1,2,3\} & \{2,3\} & \{3\} & \{\} & \{1,3\} & \{1\} \\
\{1,2,3\} & \{1\} & \{1,3\} & \{2\} & \{2,3\} & \{1,2\} & \{\} & \{3\} \\
\{3\} & \{\} & \{2,3\} & \{1,3\} & \{1,2,3\} & \{1\} & \{2\} & \{1,2\} \\
\{\} & \{2,3\} & \{1,2\} & \{3\} & \{1\} & \{1,3\} & \{1,2,3\} & \{2\} \\
\{1,3\} & \{3\} & \{\} & \{1\} & \{2\} & \{1,2,3\} & \{1,2\} & \{2,3\} \\
\{1\} & \{1,3\} & \{3\} & \{1,2,3\} & \{1,2\} & \{2\} & \{2,3\} & \{\} \\
\{2\} & \{1,2,3\} & \{1\} & \{1,2\} & \{\} & \{2,3\} & \{3\} & \{1,3\}
\end{bmatrix}$$

$$U(16)$$

$$\begin{bmatrix}
-3 & -5 & -7 & 1 & 5 & 3 & -1 & 7 \\
-5 & -1 & 7 & -7 & 1 & -3 & 3 & 5 \\
7 & 5 & -1 & -5 & -3 & -7 & 1 & 3 \\
1 & -7 & -3 & 3 & -1 & 5 & 7 & -5 \\
-7 & 7 & -5 & -3 & 3 & 1 & 5 & -1 \\
3 & -3 & 1 & 5 & 7 & -1 & -5 & -7 \\
-1 & 3 & 5 & 7 & -7 & -5 & -3 & 1 \\
5 & 1 & 3 & -1 & -5 & 7 & -7 & -3
\end{bmatrix}$$

$$\begin{bmatrix}
7 & -1 & -3 & -7 & 3 & -5 & 5 & 1 \\
-1 & 1 & -5 & 7 & -3 & 5 & -7 & 3 \\
5 & -7 & 1 & -5 & -1 & 3 & -3 & 7 \\
-7 & 7 & 3 & 5 & 1 & -3 & -5 & -1 \\
-3 & -5 & 7 & 3 & -7 & -1 & 1 & 5 \\
3 & -3 & -7 & 1 & 5 & 7 & -1 & -5 \\
-5 & 5 & -1 & -3 & 7 & 1 & 3 & -7 \\
1 & 3 & 5 & -1 & -5 & -7 & 7 & -3
\end{bmatrix}$$

$$\begin{bmatrix}
3 & 7 & 5 & -5 & -7 & -3 & -1 & 1 \\
-1 & 1 & 3 & 5 & -3 & 7 & -5 & -7 \\
-5 & -7 & -1 & 3 & 7 & 1 & 5 & -3 \\
1 & 3 & 7 & -3 & -5 & 5 & -7 & -1 \\
-3 & -5 & -7 & 1 & 3 & -1 & 7 & 5 \\
7 & 5 & -3 & -7 & -1 & -5 & 1 & 3 \\
5 & -3 & -5 & -1 & 1 & -7 & 3 & 7 \\
-7 & -1 & 1 & 7 & 5 & 3 & -3 & -5
\end{bmatrix}$$

$$\begin{bmatrix}
7 & -1 & -7 & 1 & -5 & -3 & 5 & 3 \\
1 & 7 & 5 & -7 & 3 & -5 & -3 & -1 \\
-3 & 5 & 3 & -5 & 1 & 7 & -1 & -7 \\
-7 & 1 & -3 & 5 & -1 & 3 & -5 & 7 \\
-1 & 3 & 1 & 7 & -3 & 5 & -7 & -5 \\
-5 & -3 & -1 & 3 & -7 & 1 & 7 & 5 \\
3 & -5 & 7 & -1 & 5 & -7 & 1 & -3 \\
5 & -7 & -5 & -3 & 7 & -1 & 3 & 1
\end{bmatrix}$$

$$\begin{bmatrix}
3 & -1 & 1 & 7 & -5 & -3 & 5 & -7 \\
-5 & -7 & 7 & 1 & 3 & 5 & -3 & -1 \\
7 & -3 & 3 & -5 & 1 & -7 & -1 & 5 \\
-1 & 7 & 5 & -3 & -7 & -5 & 3 & 1 \\
-3 & -5 & -1 & -7 & 5 & 1 & 7 & 3 \\
1 & 5 & -5 & 3 & 7 & -1 & -7 & -3 \\
-7 & 1 & -3 & 5 & -1 & 3 & -5 & 7 \\
5 & 3 & -7 & -1 & -3 & 7 & 1 & -5
\end{bmatrix}$$

$$\begin{bmatrix} 5 & -1 & -3 & 1 & 7 & -5 & 3 & -7 \\ 3 & -3 & 5 & -1 & 1 & 7 & -7 & -5 \\ -5 & 3 & -7 & 5 & -3 & -1 & 7 & 1 \\ 1 & -5 & 7 & -7 & 3 & 5 & -1 & -3 \\ 7 & -7 & -5 & 3 & 5 & -3 & 1 & -1 \\ -1 & 7 & 1 & -5 & -7 & 3 & -3 & 5 \\ -7 & 5 & 3 & -3 & -1 & 1 & -5 & 7 \\ -3 & 1 & -1 & 7 & -5 & -7 & 5 & 3 \end{bmatrix}$$

6.3   Chapter 4 Solutions

$$D_3$$

$$\begin{bmatrix} f & rr & rf & rrf & e & r \\ r & rf & e & f & rr & rrf \\ rr & f & r & e & rrf & rf \\ rf & r & rrf & rr & f & e \\ rrf & e & rr & r & rf & f \\ e & rrf & f & rf & r & rr \end{bmatrix} \quad \begin{bmatrix} e & r & rf & f & rrf & rr \\ rr & e & rrf & rf & f & r \\ f & rrf & rr & e & r & rf \\ rrf & rf & e & r & rr & f \\ r & rr & f & rrf & rf & e \\ rf & f & r & rr & e & rrf \end{bmatrix}$$

$$\begin{bmatrix} rf & r & rrf & f & rr & e \\ f & e & r & rr & rf & rrf \\ rr & rrf & e & rf & f & r \\ rrf & rf & rr & r & e & f \\ e & rr & f & rrf & r & rf \\ r & f & rf & e & rrf & rr \end{bmatrix} \quad \begin{bmatrix} e & rr & f & rf & rrf & r \\ rrf & rf & e & r & f & rr \\ f & r & rrf & rr & e & rf \\ rr & e & r & rrf & rf & f \\ rf & rrf & rr & f & r & e \\ r & f & rf & e & rr & rrf \end{bmatrix}$$

$$\begin{bmatrix} rr & rf & f & r & rrf & e \\ f & e & r & rr & rf & rrf \\ e & rr & rrf & rf & r & f \\ rf & r & e & rrf & f & rr \\ rrf & f & rf & e & rr & r \\ r & rrf & rr & f & e & rf \end{bmatrix} \quad \begin{bmatrix} e & r & rrf & rf & f & rr \\ rrf & rf & e & r & rr & f \\ rf & f & r & rr & rrf & e \\ rr & rrf & f & e & r & rf \\ f & e & rr & rrf & rf & r \\ r & rr & rf & f & e & rrf \end{bmatrix}$$

$$
\begin{bmatrix}
rr & f & r & rf & rrf & e \\
r & rr & f & e & rf & rrf \\
rf & rrf & e & rr & f & r \\
rrf & e & rf & f & r & rr \\
e & rf & rrf & r & rr & f \\
f & r & rr & rrf & e & rf
\end{bmatrix}
\quad
\begin{bmatrix}
rrf & rf & rr & r & f & e \\
rr & f & e & rf & rrf & r \\
rf & e & f & rr & r & rrf \\
e & rrf & r & f & rr & rf \\
r & rr & rf & rrf & e & f \\
f & r & rrf & e & rf & rr
\end{bmatrix}
$$

$$
\begin{bmatrix}
rrf & r & e & f & rf & rr \\
rf & rr & r & rrf & f & e \\
f & e & rr & rf & rrf & r \\
rr & f & rf & r & e & rrf \\
r & rf & rrf & e & rr & f \\
e & rrf & f & rr & r & rf
\end{bmatrix}
\quad
\begin{bmatrix}
e & r & f & rr & rf & rrf \\
r & e & rf & rrf & f & rr \\
f & rf & rr & r & rrf & e \\
rr & rrf & r & rf & e & f \\
rf & f & rrf & e & rr & r \\
rrf & rr & e & f & r & rf
\end{bmatrix}
$$

$$
\begin{bmatrix}
rf & rr & rrf & e & r & f \\
rr & f & e & r & rf & rrf \\
e & r & rr & f & rrf & rf \\
r & rf & f & rrf & e & rr \\
f & rrf & r & rf & rr & e \\
rrf & e & rf & rr & f & r
\end{bmatrix}
\quad
\begin{bmatrix}
f & e & rrf & rr & r & rf \\
e & f & r & rf & rrf & rr \\
rrf & r & rf & e & rr & f \\
rf & rr & f & r & e & rrf \\
rr & rf & e & rrf & f & r \\
r & rrf & rr & f & rf & e
\end{bmatrix}
$$

$$D_4$$

$$
\begin{bmatrix}
r & f & rf & rrrf & e & rrr & rrf & rr \\
rrr & rrf & rrrf & f & r & rr & e & rf \\
rr & e & f & rrf & rrr & rf & r & rrrf \\
rf & r & rrf & e & rr & rrrf & rrr & f \\
rrrf & rrr & e & r & rf & f & rr & rrf \\
f & rr & r & rrr & rrrf & rrf & rf & e \\
rrf & rf & rrr & rr & f & e & rrrf & r \\
e & rrrf & rr & rf & rrf & r & f & rrr
\end{bmatrix}
$$

$$
\begin{bmatrix}
rf & rr & rrrf & e & r & rrr & rrf & f \\
e & rrrf & rrr & rr & f & r & rf & rrf \\
rrf & e & rr & rf & rrr & rrrf & f & r \\
rrr & f & rrf & r & e & rf & rrrf & rr \\
rr & rrr & r & rrrf & rrf & f & e & rf \\
r & rrf & rf & f & rr & e & rrr & rrrf \\
rrrf & r & f & rrr & rf & rrf & rr & e \\
f & rf & e & rrf & rrrf & rr & r & rrr
\end{bmatrix}
$$

$$\begin{bmatrix} f & r & rrf & e & rrr & rr & rrrf & rf \\ rrf & rrrf & rrr & r & rr & e & rf & f \\ r & rrr & rrrf & rrf & rf & f & rr & e \\ rrrf & rr & rf & rrr & f & rrf & e & r \\ rr & f & e & rf & r & rrrf & rrf & rrr \\ rf & e & f & rr & rrf & rrr & r & rrrf \\ e & rrf & r & f & rrrf & rf & rrr & rr \\ rrr & rf & rr & rrrf & e & r & f & rrf \end{bmatrix}$$

$$\begin{bmatrix} rf & f & rrf & rr & rrr & rrrf & e & r \\ rrrf & rrr & f & rrf & rr & r & rf & e \\ rrf & rf & e & r & rrrf & f & rr & rrr \\ f & rrrf & rf & e & r & rrr & rrf & rr \\ e & rrf & rr & rrr & f & rf & r & rrrf \\ rrr & r & rrrf & rf & e & rr & f & rrf \\ r & rr & rrr & f & rrf & e & rrrf & rf \\ rr & e & r & rrrf & rf & rrf & rrr & f \end{bmatrix}$$

$$\begin{bmatrix} f & rrr & rrf & rr & rf & e & r & rrrf \\ rr & e & rf & r & f & rrrf & rrr & rrf \\ rf & r & rrrf & f & rrf & rrr & rr & e \\ rrf & rr & e & rf & rrrf & r & f & rrr \\ rrr & rrf & rr & e & r & rf & rrrf & f \\ e & rf & r & rrrf & rrr & f & rrf & rr \\ rrrf & f & rrr & rrf & e & rr & rf & r \\ r & rrrf & f & rrr & rr & rrf & e & rf \end{bmatrix}$$

$$\begin{bmatrix} rrr & rrf & r & rrrf & f & rf & e & rr \\ rrrf & r & f & rf & e & rrf & rr & rrr \\ r & rr & rrr & f & rrrf & e & rf & rrf \\ rf & f & e & rrf & rr & r & rrr & rrrf \\ f & rrr & rrrf & e & rf & rr & rrf & r \\ rr & rf & rrf & rrr & r & rrrf & f & e \\ e & rrrf & rf & rr & rrf & rrr & r & f \\ rrf & e & rr & r & rrr & f & rrrf & rf \end{bmatrix}$$

$$S_3$$

$$
\begin{bmatrix}
(13) & (23) & (123) & (12) & (132) & e \\
(132) & e & (23) & (13) & (12) & (123) \\
(23) & (13) & (12) & (123) & e & (132) \\
e & (132) & (13) & (23) & (123) & (12) \\
(12) & (123) & e & (132) & (13) & (23) \\
(123) & (12) & (132) & e & (23) & (13)
\end{bmatrix}
$$

$$
\begin{bmatrix}
(23) & (123) & (13) & e & (132) & (12) \\
e & (13) & (123) & (23) & (12) & (132) \\
(123) & (12) & (132) & (13) & (23) & e \\
(12) & e & (23) & (132) & (123) & (13) \\
(13) & (132) & (12) & (123) & e & (23) \\
(132) & (23) & e & (12) & (13) & (123)
\end{bmatrix}
$$

$$
\begin{bmatrix}
e & (23) & (12) & (132) & (123) & (13) \\
(13) & (12) & e & (23) & (132) & (123) \\
(132) & (13) & (123) & e & (12) & (23) \\
(12) & (132) & (23) & (123) & (13) & e \\
(123) & e & (13) & (12) & (23) & (132) \\
(23) & (123) & (132) & (13) & e & (12)
\end{bmatrix}
$$

$$
\begin{bmatrix}
(23) & e & (123) & (12) & (132) & (13) \\
(13) & (23) & (12) & (132) & (123) & e \\
e & (13) & (132) & (123) & (12) & (23) \\
(132) & (12) & e & (23) & (13) & (123) \\
(123) & (132) & (23) & (13) & e & (12) \\
(12) & (123) & (13) & e & (23) & (132)
\end{bmatrix}
$$

$$
\begin{bmatrix}
e & (12) & (23) & (13) & (132) & (123) \\
(123) & e & (13) & (12) & (23) & (132) \\
(13) & (23) & (123) & (132) & e & (12) \\
(12) & (13) & (132) & (23) & (123) & e \\
(23) & (132) & e & (123) & (12) & (13) \\
(132) & (123) & (12) & e & (13) & (23)
\end{bmatrix}
$$

$$\begin{bmatrix} (123) & (23) & (13) & (132) & e & (12) \\ (132) & e & (12) & (23) & (13) & (123) \\ (13) & (123) & (23) & (12) & (132) & e \\ e & (12) & (132) & (13) & (123) & (23) \\ (12) & (132) & e & (123) & (23) & (13) \\ (23) & (13) & (123) & e & (12) & (132) \end{bmatrix}$$

$$Q$$

$$\begin{bmatrix} -1 & k & i & 1 & -i & j & -k & -j \\ -k & j & -j & -i & 1 & k & -1 & i \\ k & -j & -i & -k & -1 & i & j & 1 \\ 1 & -k & k & i & -j & -1 & -i & j \\ -j & 1 & -1 & j & k & -i & i & -k \\ -i & -1 & j & -j & i & -k & 1 & k \\ i & -i & -k & k & j & 1 & -j & -1 \\ j & i & 1 & -1 & -k & -j & k & -i \end{bmatrix}$$

$$\begin{bmatrix} -k & k & j & -i & -1 & i & -j & 1 \\ 1 & -1 & -j & j & -k & -i & i & k \\ -i & -j & k & 1 & i & -k & -1 & j \\ i & j & 1 & -k & -j & -1 & k & -i \\ k & -k & i & -j & 1 & j & -i & -1 \\ j & i & -1 & k & -i & 1 & -k & -j \\ -1 & 1 & -i & i & k & -j & j & -k \\ -j & -i & -k & -1 & j & k & 1 & i \end{bmatrix}$$

$$\begin{bmatrix} 1 & -i & -j & -k & -1 & k & j & i \\ -i & k & -k & 1 & i & j & -1 & -j \\ -1 & i & k & j & 1 & -j & -k & -i \\ j & -1 & -i & k & -k & i & -j & 1 \\ -k & 1 & i & -j & j & -i & k & -1 \\ -j & -k & -1 & i & k & 1 & -i & j \\ i & -j & j & -1 & -i & -k & 1 & k \\ k & j & 1 & -i & -j & -1 & i & -k \end{bmatrix}$$

$$\begin{bmatrix} -1 & k & j & i & -j & -k & 1 & -i \\ k & -i & i & -1 & j & -j & -k & 1 \\ -j & j & 1 & -k & -i & k & -1 & i \\ i & -1 & -j & j & -k & 1 & -i & k \\ 1 & -k & k & -i & -1 & i & j & -j \\ j & i & -k & -j & 1 & -i & k & -1 \\ -i & 1 & -1 & k & i & j & -j & -k \\ -k & -j & -i & 1 & k & -1 & i & j \end{bmatrix}$$

$$\begin{bmatrix} -1 & -i & k & -k & i & -j & 1 & j \\ -k & i & j & -j & -1 & 1 & k & -i \\ j & 1 & -k & -i & k & i & -1 & -j \\ -j & -1 & -i & 1 & -k & k & j & i \\ i & j & 1 & -1 & -i & -k & -j & k \\ k & -j & -1 & j & 1 & -i & i & -k \\ -i & k & -j & i & j & -1 & -k & 1 \\ 1 & -k & i & k & -j & j & -i & -1 \end{bmatrix}$$

$$\begin{bmatrix} k & j & -i & i & 1 & -1 & -k & -j \\ 1 & -k & -1 & j & -j & k & -i & i \\ -1 & i & -k & -j & k & -i & j & 1 \\ j & k & -j & -1 & -k & i & 1 & -i \\ -j & -i & k & -k & i & 1 & -1 & j \\ i & -1 & 1 & -i & j & -j & k & -k \\ -k & 1 & i & k & -i & j & -j & -1 \\ -i & -j & j & 1 & -1 & -k & i & k \end{bmatrix}$$

6.4   Chapter 5 Solutions

Latin Square Tricks and the Group Product of All Elements

$$\begin{bmatrix} 2 & 0 & 1 & 3 \\ 1 & 3 & 0 & 2 \\ 3 & 1 & 2 & 0 \\ 0 & 2 & 3 & 1 \end{bmatrix}$$

Squares and the Number of Possibilities for Adjacent Cells

$$\begin{bmatrix} 0 & 2 & 3 & 1 \\ 2 & 3 & 1 & 0 \\ 1 & 0 & 2 & 3 \\ 3 & 1 & 0 & 2 \end{bmatrix}$$

$$\begin{bmatrix} b & c & e & a \\ c & e & a & b \\ e & a & b & c \\ a & b & c & e \end{bmatrix}$$

Tricks With Subgroups and Cosets

$$\begin{bmatrix} 3 & 1 & 2 & 0 \\ 0 & 3 & 1 & 2 \\ 2 & 0 & 3 & 1 \\ 1 & 2 & 0 & 3 \end{bmatrix}$$

$$\begin{bmatrix} a & b & c & e \\ c & e & a & b \\ b & c & e & a \\ e & a & b & c \end{bmatrix}$$

$$\begin{bmatrix} J & J & H & H \\ H & H & J & J \\ J & H & H & J \\ H & J & J & H \end{bmatrix}$$

Using Cosets with Non-abelian Groups

$$\begin{bmatrix} r & rf & f & rrf & e & rr \\ rr & e & rf & r & f & rrf \\ f & r & rrf & e & rr & rf \\ rf & rr & r & f & rrf & e \\ e & rrf & rr & rf & r & f \\ rrf & f & e & rr & rf & r \end{bmatrix}$$

References

1. Alperin, J. L., & Bell, R. B. (1995). *Groups and representations* (196 pp.). New York, NY: Springer.
2. Armstrong, M. A. (1988). *Groups and symmetry* (198 pp.). New York, NY: Springer.
3. Bannon, T. (2015). The origin of quaternions. *The College Mathematics Journal, 46*(1), 43–50.
4. Bell, E. T. (1937). *Men of mathematics* (592 pp.). New York: Simon & Schuster.
5. Davvaz, B. (2022). *A first course in group theory* (291 pp.). Springer Singapore.
6. Fraleigh, J. B. (2002). *A first course in abstract algebra* (7th ed., 544 pp.). Pearson.
7. Gallian, J. (2012). *Contemporary abstract algebra* (8th ed., 656 pp.). Cengage Learning.
8. Gray, J. (2018). *A history of abstract algebra: From algebraic equations to modern algebra.* Springer Nature Switzerland.
9. Hankins, T. L. (2004). *Sir William Rowan Hamilton.* Johns Hopkins University Press.
10. Hayes, B. (2006). Computing science: Gauss's day of reckoning. *American Scientist, 94*(3), 200–205.
11. Klein, F. (2007). *Lectures on the icosahedron and the solution of the fifth degree.* Cosimo, Inc.
12. Kleiner, I. (1986). The evolution of group theory: A brief survey. *Mathematics Magazine, 59*(4), 195–215 (21 pages).
13. Krantz, S. G. (2000). *Handbook of typography for mathematical sciences* (192 pp.). Chapman and Hall.
14. McKay, B. D., & Wanless, I. M. (2005). On the number of Latin squares. *Annals of Combinatorics, 9*, 335. https://doi.org/10.1007/s00026-005-0261-7
15. Nacin, D. (2019). Puzzles, parity maps, and plenty of solutions. In J. Beineke & J. Rosenhouse (Eds.), *From the mathematics of various entertaining subjects: Volume 3: The magic of mathematics* (pp. 222–248). Princeton University Press.
16. Rosen, K. (2010). *Elementary number theory and its application* (6th ed., 752 pp.). Pearson.
17. Rosen, M. I. (1995). Niels Hendrik Abel and equations of the fifth degree. *American Mathematical Monthly, 102*, 495–505.
18. Pengelley, D. J. (2005). Arthur Cayley and the first paper on group theory. In A. Shell-Gellasch & D. Jardine (Eds.), *From calculus to computers using the last 200 years of mathematics history in the classroom* (pp. 3–8). Cambridge University Press.
19. Rothman, T. (1982). The short life of Évariste Galois. *Scientific American, 246*(4), 136–149.
20. Rotman, J. J. (1994). *An introduction to the theory of groups* (517 pp.). Springer New York.
21. van der Waerden, B. L. (1976). Hamilton's discovery of quaternions. *Mathematics Magazine, 49*(5), 227–234.

Index

© The Author(s), under exclusive license to Springer Nature Switzerland AG 2024   163
D. Nacin, *The Group Theory Puzzle Book*, SUMS Readings,
https://doi.org/10.1007/978-3-031-70091-0